Meet the editor

Professor Dr. Eng. Lucian Mihet-Popa received his Habilitation (2015) and Ph.D. (2002) in Electrical Engineering, a master's degree (2000) in Electric Drives and Power Electronics, and a bachelor's degree (1999) in Electrical Engineering, all from the Politehnica University of Timisoara, Romania. He has been a Full Professor of Smart Energy Technology, at Oestfold University College, Norway, since 2016. He is also the founder and head of the university's research group on intelligent control of energy conversion and storage systems and one of the coordinators of the master's program in green energy technology. He has previously worked with Danish Technical University and Aalborg University, Denmark; Siegen University, Germany; and the Technical University of Timisoara. Dr. Mihet has published more than 180 papers in national and international journals and conference proceedings and 15 books. He has been granted more than twenty international grants/projects, such as FP7, EEA, and Horizon 2020, as well as more than ten national research grants.

Contents

Preface

Over the past decade, with rapidly evolving new technologies, and increased usage of digital information, monitoring, and control, the smart grid concept has been promoted all over the world, attracting huge interest. A smart grid is an intelligent electricity network that uses digital and other advanced technologies, such as cyber security communication or Internet of Energy technologies, that can be integrated in a cost-efficient manner. It can monitor and control the generation and transport of electricity to the end user/consumer in an efficient, economic, and secure way to ensure a sustainable power system with low losses and high levels of quality.

In the future, smart grids with energy management systems will include more distributed generation systems, such as renewable energy sources, as well as power electronics converters, energy storage technologies, advanced communications, and smart homes. Smart grids and electric vehicles (EVs) are two examples of fast-growing technologies that would greatly benefit from the development of advanced infrastructure and components. On the edge of the smart grid infrastructure, where consumers can produce energy from several local and distributed generation units, the need for energy management tools and control algorithms becomes an important option. To address these issues, many papers, research reports and books have proposed numerous smart grid concepts, software architectures, and simulation models/tools.

This book presents a comprehensive overview of smart grids, with information on emerging grid structures and future vision in the development and deployment of smart grids. Advancements in smart grids include advanced smart grid architectures with smart power systems containing communication and information technologies, renewable energy integration with power electronic converters and energy storage systems suitable for smart grid applications, and energy management strategies, including demand-side management (DSM) and demand response (DR). The book also discusses emerging applications, including EVs and charging stations, as well as big data management, cloud computing, and artificial intelligence.

The book contains six chapters organized into three sections. Section 1, "AC-DC Smart Hybrid Microgrids: Modeling, Control and Applications", contains two chapters. Chapter 1, "A Review on the Driving Forces, Challenges, and Applications of AC/DC Hybrid Smart Microgrids", discusses the advantages and disadvantages of AC/DC hybrid microgrids and analyzes potential applications that would benefit from such infrastructures. It also discusses the most significant efforts and requirements for the constitution of a solid regulatory framework for AC/DC hybrid grids. Chapter 2, "Hierarchical Control of an Islanded AC Micro Grid Using FS-MPC and an EMS", deals with the modelling of DER components in an AC microgrid, with each component using Finite Set-Model Predictive Control (FS-MPC) for controlling their power electronic converters. A supervisory controller for the energy management system of the microgrid to operate in different power flows through the proposed control algorithm has also been designed.

Section 2, "Smart Distribution Systems: Methodologies, Real-time Platforms and Testing Methods", also contains two chapters. Chapter 3, "Network Reconfiguration and Reactive Power Compensation Dispatch in Smart Distribution Systems", presents two methodologies for solving the reconfiguration and reactive power compensation dispatch in a smart distribution network. Both methodologies are formulated as two-stage solve reconfiguration and reactive power compensation problems with the characteristic of having acceptable computational efficiency and loss reduction close to the optimal solution. Chapter 4, "Advances and Prospects in Distributed Generation Sources Digital Twins Design", points out the most important digital twin types, highlighting their advantages and disadvantages as well as their applications. In this sense, it could be regarded as a set of guidelines on how to choose appropriate digital twin types and digital twinning platforms for some research and engineering problems.

Section 3, "Energy Storage Systems and Their Applications in Smart Grids", includes two chapters. Chapter 5, "Energy Storage Systems and Their Role in Smart Grids", presents the different types of storage systems and their advantages and disadvantages. It describes the main roles that energy storage systems will play in the context of smart grids. It also discusses interactions between energy storage systems and renewables. The emphasis is on the problems that these storage systems will have to deal with and the possible means that can be used for this purpose. Finally, the chapter presents the battery management system as a general concept. Chapter 6, "Segmented Coil Design Powering the Next Generation of High-efficiency Robust Micro-implants", introduces unique optimisation metrics for designing efficient transmit and receive coils for near-field magnetics-based charging solutions. It elaborates on how the metrics need to be altered depending on the regulatory limits and discusses the impact of body tissue loading on transmit and receive coil performance using circuit analysis. The chapter also introduces a novel "segmented" transmit coil arrangement and compares the near fields with and without tissue loading to show that segmented coils offer a significant improvement in the performance and robustness of a wireless power transfer system.

I would like to thank all authors and contributors of the individual chapters for their efforts and for the quality of the research material submitted. I would also like to thank Author Service Manager Dolores Kuzelj at IntechOpen for their support during the publishing process.

Lucian Mihet-Popa
Faculty of Information Technology, Engineering and Economics,
Østfold University College,
Fredrikstad, Norway

Section 1

AC-DC Smart Hybrid Microgrids: Modeling, Control and Applications

Chapter 1

A Review on the Driving Forces, Challenges, and Applications of AC/DC Hybrid Smart Microgrids

Maria Fotopoulou, Dimitrios Rakopoulos, Fotis Stergiopoulos and Spyros Voutetakis

Abstract

The majority of Medium Voltage (MV) and Low Voltage (LV) power systems are based on and operate using Alternating Current (AC) infrastructures. Yet, modern energy market needs, which promote more decentralized concepts with a high Renewable Energy Sources (RES) penetration rate and storage integration, bring Direct Current (DC) to the forefront. In this sense, AC/DC hybrid smart microgrids constitute a newly-introduced research field with a variety of potential applications that combine the benefits of both AC and DC systems. The purpose of this chapter is to review the advantages and disadvantages of AC/DC hybrid grids and analyze potential applications that would benefit from such infrastructures. Also, the most significant efforts and requirements for the constitution of a solid regulatory framework for AC/DC hybrid grids are presented, to pave the way towards their wider adoption by the market.

Keywords: AC/DC hybrid microgrid, microgrid applications, medium voltage, low voltage, framework

1. Introduction

Electrical grids, from the early stages of their implementation up until the past few decades, have been traditionally based on the energy generated from fossil fuels, such as coal, oil and natural gas [1]. As a result, and taking into account the technological expertise and the available technology at the time they were developed, most architectures were designed to be centralized. Moreover, the overall energy system architecture had a unidirectional approach, from centralized units of production to dispersed customers based on Alternating Current (AC) transmission and distribution networks, a process that has not been friendly to the environment [2].

Yet, the modern socioeconomic requirements and the need to take effective actions for the protection of the environment are challenging the traditional approach of electrical grid architectures [3]. The need for sustainability in the energy sector,

IntechOpen

underpinned by the recent technological developments, has led to the development of distributed, environmental-friendly and predominantly DC-based power supply system's [4]. Such systems typically include photovoltaic (PV) panels, Battery Energy Storage Systems (BESS), fuel cells, etc. [5]. This type of Distributed Energy Resources (DER) is most efficiently incorporated in smart microgrids [6]. Smart microgrids constitute advanced architectures, the key elements of which are smart sensors, advanced metering infrastructures, information technologies, Internet of Things (IoT), Cloud of Things and real-time communication systems. In this way, they enable the digitalization and decentralization of the grid, thus allowing for the efficient and seamless Renewable Energy Sources (RES) integration, the management of multiple distributed power supply units, the bidirectional power flow and a variety of grid-flexible services, such as black-start, island-mode operation and congestion management [7]. Furthermore, since many of the DER and some loads utilize DC power, research is oriented towards the design and development of AC/DC hybrid smart microgrids [8]. The structural differences between the traditional AC electrical grid and the AC/DC hybrid smart microgrid are presented in **Figure 1**.

This chapter aims to review the motives and applications of AC/DC hybrid smart microgrids. For this purpose, it is structured as follows: the driving forces for the development of AC/DC hybrid smart microgrids are analyzed in Section 2, their possible applications are analyzed in Section 3, the challenges regarding the regulatory framework for their wider adoption by the market are presented in Section 4, and finally conclusions are summarized in Section 5.

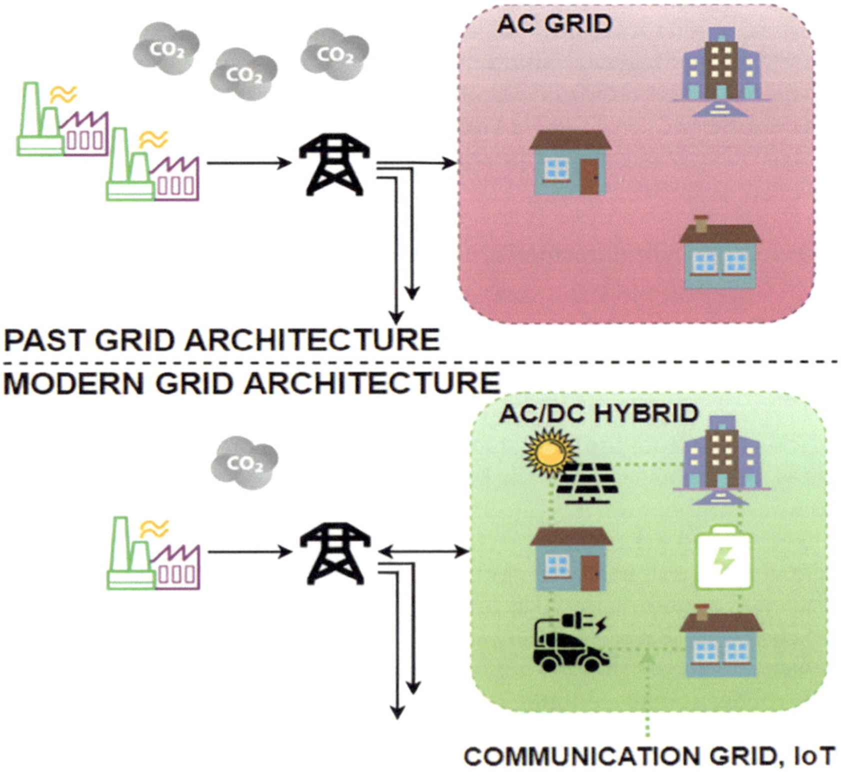

Figure 1.
Main differences between past AC and modern AC/DC hybrid smart microgrid architectures.

2. Driving forces and challenges for the development of AC/DC hybrid smart microgrids

The integration of DC systems in AC-based infrastructures provides a new framework of grid capabilities [9]. As it is foreseen, DC solutions need to "harmoniously co-exist" with the already available AC infrastructures, developing hybrid architectures in which the best result of both approaches can be achieved [10].

In this manner, AC/DC hybrid smart microgrids bring a new perspective into a variety of applications [11]. As opposed to the traditional AC infrastructures, some of their main advantages include:

- **Efficient integration of DER and reduction of primary energy consumption** [12, 13]: A majority of RES and storage systems utilized produce DC power, which would be more efficiently deployed in a DC grid instead of an AC grid, rather than having to undergo DC/AC and AC/DC conversions reducing efficiency. Typical examples are considered to be PVs, BESS, fuel cells or even EVs which may be used as additional storage, in vehicle-to-grid (V2G) mode [14]. More specifically, in a DC grid, these sources' supply is not required to be converted from DC to AC. Contrariwise, instead of DC/AC inverters, DC/DC converters need to be installed [15], as presented in **Figure 2**, which have better efficiency and a smaller size and volume. Furthermore, wherever it is necessary, bidirectional DC/DC converters can be employed, for example in the case of BESS. As a result of the above, the RES generation is exploited to the maximum, the primary energy consumption is reduced and potentially required space is reduced [16].

- **Efficient integration of DC loads** [17]: The distribution of DC power (instead of AC) to DC loads may result in energy savings from the point of view of the end-users. By eliminating the conversion from AC power to DC power with the use of the respective AC/DC inverter, as presented in **Figure 2**, the associated losses are reduced, resulting in lower electricity bills. This modification has the potential to lead to substantial cost savings especially in the case of DC loads such as computer data centers (which predominantly use DC power for their electronic equipment), EVs and EV charging stations, Light Emitting Diode (LED) equipment, etc. [18].

- **Power transmission over long distances** [19]: Considering the effectiveness of high and medium-voltage DC lines, more power may be transferred over

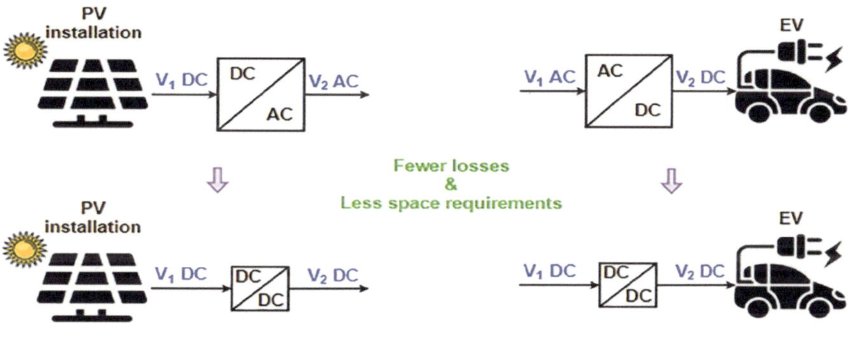

Figure 2.
Efficient integration of DC power supply and DC loads in AC/DC hybrid smart microgrids.

long distances, compared to their AC counterparts, owing to less severe stability issues. This feature makes DC integration into already existing AC grids extremely valuable for the future expansion of the grid and the connection of remote loads (such as in islands or isolated locations) or remote RES installations [20]. Major technological developments have enabled the increase of DC operating voltage levels in the order of kV, allowing efficient and reliable power transfer even for distances in the order of several thousands of kilometers.

- **Power quality enhancement** [21]: DC connections can enhance the power quality of weak grids, as they provide a type of isolation "firewall" that prevents the propagation of disturbances, as depicted in **Figure 3**. In particular, the conversion from AC to DC power decouples the AC part of the grid, so that the remaining infrastructure can cope with undesirable resonances which would otherwise impose a threat in its stability and robustness. A typical example of such cases is AC harmonic oscillations, frequency instability and low inertia, which AC grids may face, due to the existence of inductive and capacitive elements, etc. In this sense, AC/DC hybrid smart microgrids present a clear advantage, especially when it comes to the connection between two AC grids with the use of DC subsystems. Furthermore, the DC connection enables the effective integration of AC systems operating with different voltage/frequency levels.

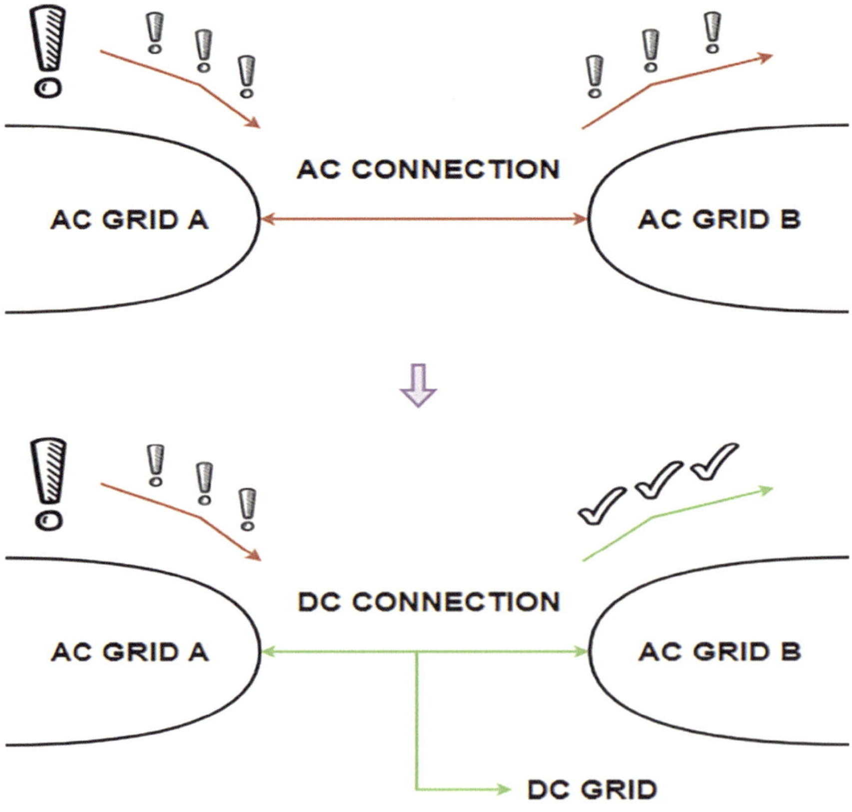

Figure 3.
DC connections prohibit the propagation of disturbances.

- **Reduction of visual impact** [22]: Since DC lines carry only active power and have no skin effect, less current capacity is required and the necessary distance between the conductors can be reduced. This, in combination with the fact that fewer lines are required than in the respective AC systems, results in a smaller size of DC towers. The smaller size of DC towers, compared to the size of AC equivalent structures, is considered to be an advantage of DC grids. This attribute, however minor it may seem, is quite beneficial considering overpopulated areas such as cities or places where the visual impact of the grids should be minimized, such as in tourist attractions, monument areas, preserved ecosystems, etc.

On the other hand, AC/DC hybrid smart microgrids have certain drawbacks. DC technologies and the connection between AC and DC technologies have not been thoroughly studied as the common AC grids. This is attributed to the fact that the entire concept of electrical energy production, transmission and distribution has been built on AC technology, which has provided the means to progress and develop simple and cost-effective AC equipment over the years.

In this context, the implementation of DC solutions has certain disadvantages such as the lack of specific standards [23]. For a newly-introduced system, such as the AC/DC hybrid grid to establish its case against the traditional AC "status quo", the definition of certain parameters needs to be specified. Since AC/DC applications are not as well-known and commonly used as AC applications, there is a general lack of standardized practices regarding their design and operation. This issue needs to be addressed, for the AC/DC hybrid grids to effectively enter the worldwide market.

Also, there is difficulty regarding the integration of DC systems into existing AC grids, to form AC/DC hybrid grids. More specifically, AC technologies have a simple design that has been studied and developed for many decades and is well known to grid developers and system operators. On the other hand, fewer specialists have studied DC technologies to that extent [24]. As a result, the incorporation of DC solutions to the existing grid bears difficulties in comparison with the application of AC solutions. In simple words, AC and DC systems have different starting points: the AC technology is proven and mature, whereas DC technology is in a developing process to be established, considering that power electronics converters started being utilized in the last quarter of the past century.

This is also reflected in protection and safety apparatus [25]. Once a new system is proposed, protection issues including switches, grounding and fault management systems need to be studied and established. In the case of AC/DC hybrid grids, there are protection issues that are not only related to the lack of standards but also the very nature of DC current. To be more specific, breaking a DC circuit is considered to be more difficult than the respective AC circuit because there is no natural zero crossing of the current to minimize the arc effect. For this purpose, major research efforts are carried out for the development of switchgear that can accommodate the secure disruption of DC voltages in the order of kVs, to enable the safe and reliable development of AC/DC grid infrastructures.

Furthermore, the point of common coupling between the AC and DC parts of the AC/DC hybrid smart microgrid introduces complexity to the overall architecture. Since a common AC distribution transformer is not capable of providing DC links, the AC/DC hybrid grid requires a different type of interface. Following the latest technological developments, this interface is the Solid State Transformer (SST) [26] which is an advanced power converter that can provide multiple ports, regardless of the voltage level or type. In this sense, it can be connected to the MV side of the AC distribution line and provide AC and DC connections at both MV and LV levels. This active power converter is modular, scalable and capable of providing grid-flexible services. Yet, to function properly, it requires advanced control systems [27], which take into account both AC and DC components, have high maintenance requirements and their advantages are reflected in their cost.

Nevertheless, all of these disadvantages can be overcome if particular attention is given to aspects where DC technology may have significant potential, to be firmly established, starting to build from that point forward. The increasing use and development of modern power electronic converters can significantly help for the diffusion of DC technology providing the necessary framework, backed up by the wider application of RES technologies.

Overall, the development of AC/DC hybrid smart microgrids appears to have many advantages, rendering them a key driver in paving the way towards energy efficiency, sustainability and mitigation of anthropogenic climate change. For them to be established in the wider market, the main applications that would highlight their potential need to be taken into consideration.

3. Main applications of AC/DC hybrid smart microgrids

This section aims to showcase modern examples of applications of AC/DC hybrid smart microgrids, which mostly concern buildings, public installations, remote installations, DC-based applications and transportation:

- **Buildings:** One of the most promising applications that would benefit from the AC/DC hybrid smart microgrid architecture is the building sector. Due to environmental as well as economic concerns, PV panels are commonly installed in buildings [28]. Surplus PV generation is usually stored in BESS, which can smooth the mismatch between the PV generation profile and the load profile. Since the PVs and the storage are both installed within the building, they allow the minimization of transmission losses (and thus, the minimization of primary energy consumption and associated CO_2 emissions). Both of these power supply units originally produce DC power. Also, a proportion of the overall load of the building, such as electronic appliances, DC motors, power electronics, batteries, etc., originally consumes DC power [29]. Therefore, it is evident that it would be more beneficial if at least part of the building's power supply was based on DC power distribution, as presented in **Figure 4** [30].

  Yet, it should be noted that while this approach could be easily implemented on the side of the DC power supply, it would be more difficult to implement on the DC consumption side, as most DC devices are designed to include (internally) an AC/DC converter to operate with AC grids [31]. This is a barrier that needs to be overcome by the manufacturers of these devices through the proper and widely publicized dissemination of DC capabilities. However, it should be noted that even without the proposed modification of DC loads, it would be beneficial to have a DC sub-grid (in the overall AC grid of the building) for the connections between the PVs, BESS and other DC-based power supply units.

- **District/Distribution level:** The suitability of AC/DC hybrid smart microgrids can be expanded from a single building application to a district-level application, as presented in **Figure 5** [32–34]. As in buildings, so in distribution grids, the penetration of DC-based RES and storage renders the AC/DC hybrid configuration more effective than the conventional AC one. This topic has gained much attention over the past few years at both Medium Voltage (MV) and Low Voltage (LV) levels.

- **Public installations:** A beneficiary of AC/DC hybrid grids could be various public installations. More specifically, as part of public works and services, older lighting equipment is often replaced by LED technology in most public spaces, roads and highways. Such initiatives help reduce the effect on the environment, as LEDs are more efficient than conventional lighting equipment. Since LED lights constitute DC-based technologies, they would naturally be more efficiently powered by DC lines (incorporated in the overall AC design, thus forming AC/DC hybrid grids) producing a significant economic impact, as public lighting costs are a major part of public expenditure [35, 36].

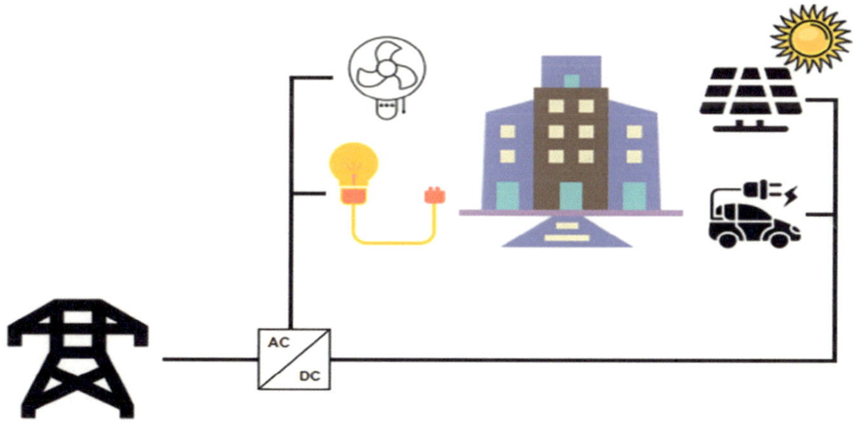

Figure 4.
Building with AC/DC hybrid smart microgrid architecture.

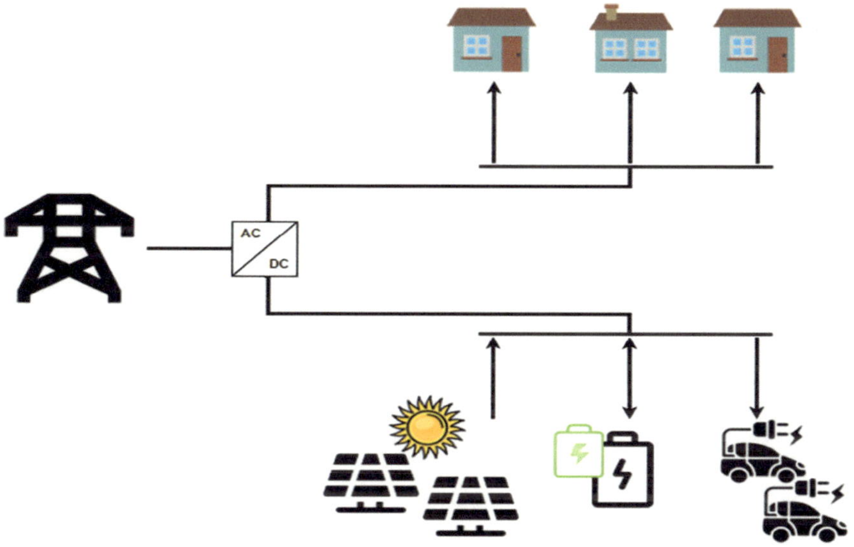

Figure 5.
District with AC/DC hybrid smart microgrid architecture.

- **Connection of remote/weak installations:** As one of the main advantages of DC lines over their AC counterparts is the capability of power transmission over long distances and the enhancement of power quality and stability, one of the key-applications of AC/DC hybrid grids is the connection to remote installations [37]. These could be small islands or any other case of distant and remote areas, which may be connected to each other or with the robust grid of the mainland through underground or underwater DC cables. Other types of locations could also be remote wind generation parks, built either on islands or offshore [38].

- **DC-based applications:** There is several DC-based applications that could benefit from a direct DC distribution. The two most modern and important are a) data centers and b) EV charging stations.

  Data centers are extremely significant facilities, whose importance is gradually increased over time, leading to the increase of required capacity for information storage. Future data centers could entail power levels up to a few MWs in order to properly function. The majority of loads in data centers are of digital nature and operate on DC power. This means that AC connections would not facilitate their development as there would be significant losses and reliability issues due to the required conversion stages. The aforementioned facts favor the adoption of AC/DC hybrid architectures in data centers [39].

  Also, as mentioned above, there is a recent, increasing need for efficient EV charging stations. In particular, when it comes to vehicles, the need to a) protect the environment from the emissions of fossil fuels, b) reduce the noise level in the urban field and c) reduce the cost of transportation, has led the car manufacturers to focus on developing and producing EVs. The sales of EVs are gradually increasing around the world and it is estimated that in the near future they shall completely replace fossil-fuel-powered vehicles [40]. EVs need charging at regular intervals and their batteries are inherently DC-power sources [41]. Therefore, charging EVs in DC-based charging stations is more effective than the respective AC alternative, which encourages the research and development towards AC/DC hybrid smart microgrids including EV charging stations [42].

  In addition to the arguments described above, it is noted that research is also oriented towards data centers and EV charging stations with PV and/or BESS installations for the purpose of reduction of a) cost, b) energy footprint and c) dependence on the main grid. These amendments furtherly favor the deployment of AC/DC hybrid grids for these applications [43].

- **Transportation:** There is a variety of applications in transport at both MV and LV levels that either already uses DC power or are prompted to do so. Typical examples include ships, urban transport and railways.

  Ships constitute a special, isolated from the main grid, application that needs large amounts of power to operate properly. Also, their design has certain limitations, due to constraints imposed by the ship's needs, including constant power availability, space and weight concerns and the presence of pulsed electric loads. DC systems, which have less volume and weight and are more appropriate for handling electronic loads (compared to their AC counterparts) are proposed to be a viable solution for ships, thus forming AC/DC architectures [44].

Urban transport vehicles and railways are one of the early adopters of DC architectures. In many cases, motors and auxiliary circuits inside urban transport vehicles use DC power. As a result, they form DC power systems, drawing power from the main AC grid of the city, through the appropriate AC/DC converters [45].

Overall, there is a variety of applications that could benefit from AC/DC hybrid smart microgrids. **Table 1** summarizes the aforementioned categories of applications, along with some of their main features, i.e., a justification for which type of architecture is suitable for each category, their voltage level and comments. It is noted that the main factors for each application are related either to the increase of RES and DC loads or to the reliability and robustness of DC connections.

Application	Reason for development of AC/DC hybrid architectures	Voltage level	Special comments
Building	Increase of building-integrated RES, BESS storage, and DC load	LV ($\leq$ 400 V)	The use of the building (residence or office) may affect the grid's design (BESS capacity, etc.)
District/Distribution level	Increase of distributed RES and storage units	LV and MV (up to several kV)	A highly DC-based district may also effectively be designed as a DC microgrid
Public installations	Increase of DC load	LV (24 V for LED lighting systems)	LED lighting is a significant part of DC public installations
Connection of remote/ weak installations	Robustness provided by DC links and efficiency of power transmission	MV or HV (depending on the application)	Underwater DC cables connect geographical islands
DC-based applications	Increase of DC load (but also possibility for PV-BESS installation)	LV (400 V, etc.)	Data centers, EV charging stations, etc.
Transportation	Reliability, size limitations for the installed components, DC load	LV and MV (usually 750 V, 1 kV, 3 kV, etc.)	Ships, railways, public transport, etc.

Table 1.
Applications of AC/DC hybrid smart microgrids [21, 46, 47].

4. Challenges regarding AC/DC hybrid smart microgrids

Several developments regarding AC/DC hybrid smart microgrids have taken place over the past few years, with fruitful results presented in the worldwide literature. Although, their effectiveness in certain applications is evident and generally accepted by the research community, there is several factors that inhibit their wide deployment, as presented in **Figure 6** [47].

First of all, these developments have been conducted separately, taking into account and focusing on the specific needs of each application, hence lacking a more general and common framework of the application. To establish their place in the market and challenge the dominance of conventional AC grids, a common legislative background for AC/DC hybrid smart microgrid solutions is considered to be a necessity [48]: it is important to establish standards upon which the architecture of AC/DC hybrid smart microgrids can be designed and implemented.

In this sense, DC compatible equipment needs to be developed by the manufacturers [21]. More specifically, one of the main reasons why such advanced grids are researched is the ascending amalgamation of DC devices in the overall load of the system. As mentioned previously, such devices include EVs, computers and other electronic devices, power electronics, DC motors, LED lights, etc. Nevertheless, currently, most of these devices are designed to be powered by AC sources. To be efficiently incorporated in AC/DC hybrid smart microgrids, they need to be designed to be powered by DC sources by incorporating a DC/DC converter [49]. For this purpose, it is imperative to establish mechanisms that promote and provide financial support to the cooperation between public and private entities, researchers and industry, allowing the development of DC-compatible equipment that is not yet available as well as suitable disconnecting and protection devices [50].

Furthermore, there is a generalized requirement for the standardization of the voltage level on hybrid grid applications, new safety regulations and suitable protection mechanisms [51–54]. More specifically, a major challenge for voltage standardization on the DC part of hybrid grids is the use of different voltage levels in distributed generation, residential, commercial and industrial demand sides. So far,

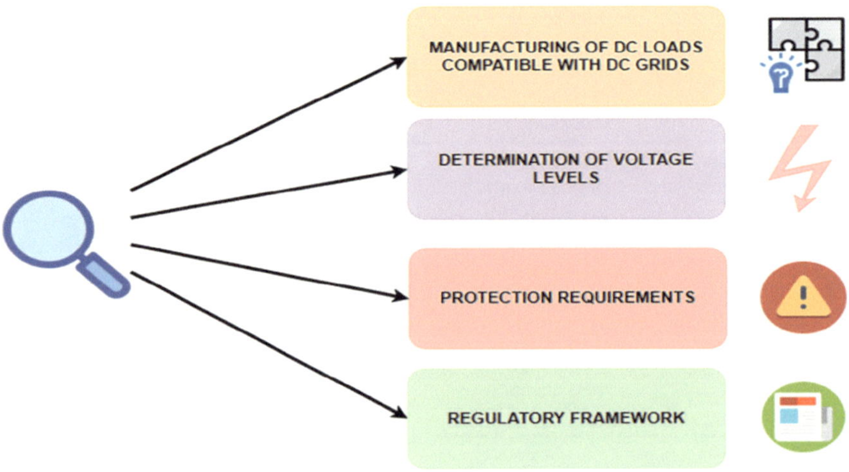

Figure 6.
Challenges for AC/DC hybrid smart microgrids.

the research community has not agreed to use a specific DC voltage level or even set clear limits between what is considered to be "low", "medium" and "high" voltage, in terms of standardization. Without voltage levels standardization it is impossible to develop appliances, equipment and devices that are directly connected to DC busses. It is inconvenient for manufacturers to design DC products capable of operating with different voltage levels. To speed up the incorporation of DC technologies in the distribution grid, voltage standardization is by far the highest priority. In this way, stakeholders, equipment manufacturers, consumers and users can be attracted to hybrid grids, increasing their readiness level.

5. Conclusions

This chapter reviewed the motives and applications of AC/DC hybrid smart microgrids. This type of grid constitutes a milestone in the evolution of electrical transmission and distribution systems, as it facilitates the efficient incorporation of the majority of RES, storage as well as DC-based loads, while also having AC connections for the service of AC generation and consumption. Indicative applications that would benefit from such architectures include buildings, data centers, EV charging stations, etc. However, the wider adoption of AC/DC hybrid smart microgrids requires a more coordinated effort in terms of the regulatory framework, so that their DC part can be as highly standardized as the AC one.

Funding

This research has received funding from the European Union's Horizon 2020 research and innovation programme, TIGON (Towards Intelligent DC-based hybrid Grids Optimizing the Network performance), under grant agreement No 957769, https://cordis.europa.eu/project/id/957769.

Conflict of interest

The authors declare no conflict of interest.

Author details

Maria Fotopoulou, Dimitrios Rakopoulos*, Fotis Stergiopoulos and Spyros Voutetakis
Centre for Research and Technology Hellas, Athens, Greece

*Address all correspondence to: rakopoulos@certh.gr

References

[1] Kaldellis JK, Zafirakis D, Kondili E. Contribution of lignite in the Greek electricity generation: Review and future prospects. Fuel. 2009;**88**:475-489

[2] Worighi I, Maach A, Hafid A, et al. Integrating renewable energy in smart grid system: Architecture, virtualization and analysis. Sustain Energy Grids Netw. 2019;**18**:100226

[3] Leonhardt R, Noble B, Poelzer G, et al. Advancing local energy transitions: A global review of government instruments supporting community energy. Energy Research and Social Science. 2022;**83**:102350

[4] Li L, Liu C, Zhang W, et al. Investment decisions in distributed renewable energy considering economic performance and life-cycle environmental impact. Computers and Industrial Engineering. 2021;**162**:107732

[5] Khezri R, Mahmoudi A, Aki H. Optimal planning of solar photovoltaic and battery storage systems for grid-connected residential sector: Review, challenges and new perspectives. Renewable and Sustainable Energy Reviews. 2022;**153**:111763

[6] Lamnatou C, Chemisana D, Cristofari C. Smart grids and smart technologies in relation to photovoltaics, storage systems, buildings and the environment. Renewable Energy. 2021;**S0960148121015883**

[7] Ben SS. Prosumer in smart grids based on intelligent edge computing: A review on Artificial Intelligence Scheduling Techniques. Ain Shams Engineering Journal. 2021;**S2090447921002409**

[8] Huang M, Zhao J, Wei Z, et al. Decentralized robust state estimation for hybrid AC/DC distribution systems with smart meters. International Journal of Electrical Power & Energy Systems. 2022;**136**:107656

[9] Unamuno E, Barrena JA. Hybrid ac/dc microgrids—Part I: Review and classification of topologies. Renewable and Sustainable Energy Reviews. 2015;**52**:1251-1259

[10] Monteiro V, Martins JS, Aparício Fernandes JC, et al. Review of a Disruptive Vision of Future Power Grids: A New Path Based on Hybrid AC/DC Grids and Solid-State Transformers. Sustainability. 2021;**13**:9423

[11] Fotopoulou M, Rakopoulos D, Trigkas D, et al. State of the Art of Low and Medium Voltage Direct Current (DC) Microgrids. Energies. 2021;**14**:5595

[12] Ene PC, Okoh CC, Okoro PA, et al. Application of smart DC-Grid for efficient use of solar photovoltaic system in driving separately excited DC motor: Dynamic performance and techno-economic assessments. Clean Eng Technol. 2021;**4**:100136

[13] Rauf S, Wahab A, Rizwan M, et al. Application of Dc-grid for Efficient use of solar PV System in Smart Grid. Procedia Comput Sci. 2016;**83**:902-906

[14] Fotopoulou M, Rakopoulos D, Blanas O. Day Ahead Optimal Dispatch Schedule in a Smart Grid Containing Distributed Energy Resources and Electric Vehicles. Sensors. 2021;**21**:7295

[15] Dhimish M, Schofield N. Single-switch boost-buck DC-DC converter for industrial fuel cell and photovoltaics applications. International Journal of Hydrogen Energy. 2021; **S0360319921041021**:1241

[16] Pourbehzadi M, Niknam T, Aghaei J, et al. Optimal operation of hybrid AC/DC microgrids under uncertainty of renewable energy resources: A comprehensive review. International Journal of Electrical Power & Energy Systems. 2019;**109**:139-159

[17] Shen L, Cheng Q, Cheng Y, et al. Hierarchical control of DC micro-grid for photovoltaic EV charging station based on flywheel and battery energy storage system. Electric Power Systems Research. 2020;**179**:106079

[18] Nandini KK, Jayalakshmi NS, Jadoun VK. An overview of DC Microgrid with DC distribution system for DC loads. Mater Today Proc. 2021;**S2214785321044485**

[19] Yamaguchi S. Asian international grid connection and potentiality of DC superconducting power transmission. Glob Energy Interconnect. 2018;**1**:9

[20] Isuru M, Hotz M, Gooi HB, et al. Network-constrained thermal unit commitment for hybrid AC/DC transmission grids under wind power uncertainty. Applied Energy. 2020;**258**:114031

[21] Kumar D, Zare F, Ghosh A. DC Microgrid Technology: System Architectures, AC Grid Interfaces, Grounding Schemes, Power Quality, Communication Networks, Applications, and Standardizations Aspects. IEEE Access. 2017;**5**:12230-12256

[22] ABB. ABB Review HVDC Special Report

[23] Van den Broeck G, Stuyts J, Driesen J. A critical review of power quality standards and definitions applied to DC microgrids. Applied Energy. 2018;**229**:281-288

[24] Dagar A, Gupta P, Niranjan V. Microgrid protection: A comprehensive review. Renewable and Sustainable Energy Reviews. 2021;**149**:111401

[25] Srivastava C, Tripathy M. DC microgrid protection issues and schemes: A critical review. Renewable and Sustainable Energy Reviews. 2021;**151**:111546

[26] Mishra DK, Ghadi MJ, Li L, et al. A review on solid-state transformer: A breakthrough technology for future smart distribution grids. International Journal of Electrical Power & Energy Systems. 2021;**133**:107255

[27] Kumar J, Agarwal A, Agarwal V. A review on overall control of DC microgrids. J Energy Storage. 2019;**21**:113-138

[28] Ceran B, Jurasz J, Mielcarek A, et al. PV systems integrated with commercial buildings for local and national peak load shaving in Poland. Journal of Cleaner Production. 2021;**322**:129076

[29] Falaki F, Merabtine A, Martouzet D. A Spatio-Temporal Analysis of electric appliance end-use demand in the residential sector: Case study of Tours (France). Sustainable Cities and Society. 2021;**65**:102635

[30] Yu H, Niu S, Zhang Y, et al. An integrated and reconfigurable hybrid AC/DC microgrid architecture with autonomous power flow control for nearly/net zero energy buildings. Applied Energy. 2020;**263**:114610

[31] Turksoy O, Yilmaz U, Teke A. Efficient AC-DC power factor corrected boost converter design for battery charger in electric vehicles. Energy. 2021;**221**:119765

[32] Agbemuko AJ, Domínguez-García JL, Gomis-Bellmunt O. Impedance-based

modelling of hybrid AC/DC grids with synchronous generator for interaction study and dynamic improvement. Electric Power Systems Research. 2020;**179**:106086

[33] Adi FS, Song H, Kim J-S. Interlink Converter Controller Design based on System Identification of DC Sub-Grid Model in Hybrid AC/DC Microgrid. IFAC-Pap. 2019;**52**:45-50

[34] Ortiz L, Orizondo R, Águila A, et al. Hybrid AC/DC microgrid test system simulation: Grid-connected mode. Heliyon. 2019;**5**:e02862

[35] Quintana PJ, Huerta N, Rico-Secades M, et al. Control of public dc street/road lighting microgrids with microgeneration and storage capability based on a power-line signaling dependent droop. Guanajuato, Mexico: 2016 13th International Conference on Power Electronics (CIEP), IEEE; 2016. pp. 98-103

[36] Liang D, Zou J, Wang Z, et al. Research on DC Vacuum Switch of Micro-Grid in Road Lighting. Xi'an: 2018 2nd IEEE Advanced Information Management, Communicates, Electronic and Automation Control Conference (IMCEC), IEEE; 2018. pp. 242-246

[37] General Electric Company. High Voltage Direct Current Systems. Powers Ferry Road Atlanta, GA, USA; 2018. Available from: https://www.gegridsolutions.com/products/brochures/powerd_vtf/hvdc-systems_gea-31971_hr.pdf [Accessed: 14 April 2021]

[38] Dhua D, Huang S, Wu Q. Optimal power flow modelling and analysis of hybrid AC-DC grids with offshore wind power plant. Energy Procedia. 2017;**141**:572-579

[39] AlLee G, Tschudi W. Edison Redux: 380 Vdc Brings Reliability and Efficiency to Sustainable Data Centers. IEEE Power Energy Mag. 2012;**10**:50-59

[40] Rietmann N, Hügler B, Lieven T. Forecasting the trajectory of electric vehicle sales and the consequences for worldwide CO2 emissions. Journal of Cleaner Production. 2020;**261**:121038

[41] Fachrizal R, Shepero M, Åberg M, et al. Optimal PV-EV sizing at solar powered workplace charging stations with smart charging schemes considering self-consumption and self-sufficiency balance. Applied Energy. 2021;**118139**

[42] Sharma G, Sood VK, Alam MS, et al. Comparison of common DC and AC bus architectures for EV fast charging stations and impact on power quality. eTransportation. 2020;**5**:100066

[43] Kumar V, Teja VR, Singh M, et al. PV Based Off-Grid Charging Station for Electric Vehicle. IFAC-Pap. 2019;**52**:276-281

[44] Hardan F, Norman R. Balancing loads of rotating generators utilizing VSC direct power controllers in a ship AC/DC smartgrid. Electric Power Systems Research. 2020;**182**:106200

[45] Verdicchio A, Ladoux P, Caron H, et al. New Medium-Voltage DC Railway Electrification System. IEEE Trans Transp Electrification. 2018;**4**:591-604

[46] ABB. Medium Voltage Products, Technical Application Papers No. 24-Medium Voltage Direct Current Applications. Dalmine: ABB; 2017

[47] Dragicevic T, Lu X, Vasquez JC, et al. DC Microgrids—Part II: A Review of Power Architectures, Applications, and Standardization Issues. IEEE Transactions on Power Electronics. 2016;**31**:3528-3549

[48] Chandra A, Singh GK, Pant V. Protection techniques for DC microgrid-A review. Electric Power Systems Research. 2020;**187**:106439

[49] Al-Ismail FS. DC microgrid planning, operation, and control: A comprehensive review. IEEE Access. 2021;**9**:36154-36172

[50] Mirsaeidi S, Dong X, Said DM. Towards hybrid AC/DC microgrids: Critical analysis and classification of protection strategies. Renewable and Sustainable Energy Reviews. 2018;**90**:97-103

[51] IEEE. Standard for Interconnection and Interoperability of Distributed Energy Resources with Associated Electric Power Systems Interfaces. IEEE; 2018 . DOI: 10.1109/IEEESTD.2018.8332112

[52] IEEE. Recommended Practice for the Design of DC Power Systems for Stationary Applications. IEEE; 2020. DOI: 10.1109/IEEESTD.2020.9206101

[53] IEEE. Standard for DC (3200 V and below) Power Circuit Breakers Used in Enclosures. IEEE; 2015. DOI: 10.1109/IEEESTD.2015.7118113

[54] IEEE Std 1709™-2010. IEEE Recommended Practice for 1 kV to 35 kV Medium-Voltage DC Power Systems on Ships. New York, NY, USA: IEEE; 2010. p. 54

Chapter 2

Hierarchical Control of an Islanded AC Micro Grid Using FS-MPC and an EMS

Andreas Pedersen, Ibrahim Ahmed and Lucian Mihet-Popa

Abstract

Microgrids and distributed energy resources (DERs) are gaining popularity owing to their efficient operation, autonomy, and dependability. Microgrids provide several new opportunities, one of which is the ability to deliver electricity continuously, even in the event of a grid failure. This chapter will first describe the modeling of DER components in a microgrid, with each component using Finite Set-Model Predictive Control (FS-MPC) for controlling the inverters to be robust, to have a fast response, to account for multiple objectives, and to eliminate manual tuning. In addition, droop control will be used to provide a voltage reference for the FS-MPC. The PV-inverter will operate as a grid- forming inverter, while the other inverters will serve as grid-feeding inverters. The proposed inverter models are validated using simulations. The microgrid has been modeled using MATLAB-Simulink software package. A supervisory controller for energy management system of the microgrid to operate in different power flows through the proposed control algorithm has also been designed. The simulation results show the effectiveness and robustness of the proposed controller during dynamic performance and transients, and the developed energy management system algorithm successfully controlled the power flow to ensure continuous power delivery to the load under all circumstances.

Keywords: AC microgrid (MG), droop control, finite-set model predictive control (FS-MPC), hierarchical control, islanded operation, energy management system (EMS), energy storage systems (ESSs)

1. Introduction

Today's society needs a dependable supply of electricity to consumers and prosumers, with high power quality. As a result of the continual adoption of novel technologies, the structure of the power grid in many countries is continuously developing, posing difficulties with energy flow changes, capacity limits, and high investment expenditures to update the power grid. For many years, power grids have been digitalized to allow centralized monitoring and administration of the power network, which is a result of the emergence of new technologies [1–6]. This "smart" digitalized grid has been a reality for the high-voltage section of the power system for a considerable amount of time, and the modernization of the low-voltage distribution sector is also in progress [7–9]. In the next years, utility grids will depend more and more on renewable energy, and users will reap the benefits of smart technologies such as electric car chargers and smart meters [10–13].

This gives an opportunity to further digitize the distribution (low-voltage) sector of the grid. One method to do this is by establishing a microgrid. In the event of maintenance or grid failure, microgrids should be able to function independently of the utility grid [14].

Hierarchical control structures consist of a primary control layer that has a quick response in milliseconds, a secondary control layer that is used to reduce steady-state errors and acts in a couple of seconds, and finally, a tertiary control layer that controls the active and reactive energy flow within the microgrid by sending power references either manually by the grid operator or automatically by an Energy Management System (EMS) that balances the net power within the microgrid [1–3].

Due to the intermittent nature of RES, it is required to incorporate a backup power source such as a battery storage system, and perhaps an additional fuel-based power source, so that the microgrid may continue to run even if the battery is depleted or the maximum discharge current is reached [15–17].

The authors in [18] proposed a MPC strategy developed in Python to optimize energy production and load management for interactive buildings integrated PV & BESS (battery energy storage system). The forecasting method used in the study involves Weighted Moving Average (WMA) combined with Trigg's tracking signal and adjustment formulas. This method includes sensitivity parameters and thresholds that allow a stricter or looser approach to be taken in forecasting time series. The proposed method adjusts the forecasted values for the rest of the planning horizon based on the deviation detected between forecasted and real-time series. The adjustment formula of a building's PV production is different from the adjustment formula of its load since the production of PV is more predictable than the load of a building, especially when it comes to residential loads.

In these research papers [19–22] the performance of the MPC design procedure for DC-DC and DC-AC converters applied to a PV system was analyzed. The authors in [19] presented a continuous control set MPC designed for a DC-DC buck converter used in a MPPT of a PV module, while in [20] an adaptive MPC for current sensorless MPPT in PV systems was evaluated. The papers [21, 22] address the optimal control problems of a grid-connected PV inverter system MPC-based MPPT method. The steady-state and dynamic performance of the MPC-based system are verified and compared with traditional controllers.

Furthermore, authors in [23] presented an examination of a predictive control method designed to prevent imbalances between the load demand and the generation capacity in an islanded microgrid. The Nonlinear Model Predictive Control (NMPC) is utilized to calculate load shedding and manage energy from batteries within an optimization framework. This results in the establishment of an optimal control problem that integrates all the microgrid's operating conditions, including load priorities for disconnection, and charging and discharging cycles of batteries. Simulation results of the microgrid's performance with and without the Microgrid Central Controller (MGCC) were compared. The results demonstrate that the control strategy can improve the reliability of the microgrid when operating in islanded mode, as the control strategy can maintain the voltage and frequency of the microgrid within safe limits and achieve a correct balance between generated power and load demand.

The motivation to carry out this study is the growing interest of RES based DG units & ESSs. This chapter is focused on modeling and simulation of an AC microgrid, developed in MATLAB-Simulink environment, which consists of a hydrogen fuel cell, a solar farm, a wind turbine, and a utility grid [24–32]. Verification by simulations with a hierarchical control structure to operate the AC microgrid in islanded mode has been performed. The primary control mainly consists of FS-MPC, where the solar farm inverter is modeled as a grid-forming inverter [29, 30, 33], and FS-MPC, which was modeled as described in [34–38], is shown to be highly robust in a variety of different scenarios. The EMS is created to protect the battery' SOC, and the maximum charging/discharging current from being reached while keeping the power balance stable in the microgrid. The constraints of operating the AC microgrid in islanded operation are the maximum discharging off the battery and fuel cell, and the stochastic RES. The novelty of this research chapter is the use of FS-MPC in the primary control with a new EMS algorithm that is highly robust during islanded operation.

The main contributions of this chapter can be summarized as follows:

1. Modeling and simulation of DER components in an AC Microgrid, including the primary and secondary controllers.

2. The primary controllers are designed to be robust and resilient based on finite-set model predictive control, which is capable of handling sudden changes in the load demand.

3. An algorithm for EMS is developed for managing the energy flow if the microgrid is completely detached from the grid and operating in islanded mode.

4. Various scenarios are developed to test and prove the robustness of the controllers.

The chapter is organized as follows. Section 2 describes the modeling of DER components in the AC microgrid, including the design of the LCL grid inverter filter, while in Section 3 the proposed EMS strategy is described and validated by simulations, using many different scenarios. The conclusion section summarizes the main outcomes of the paper.

2. Modeling of DER components

This section describes how we created and developed detailed models for many microgrid components that make up the proposed microgrid. In the following sections, each component, including equations, parameters, and other design factors, will be thoroughly examined. The MATLAB-Simulink software program is used to implement all simulation models. The various component models are well-known and available in the literature, however the parameters, filter designs, and converter designs have been adjusted and chosen to match the requirements of the proposed microgrid model.

The general structure of the proposed microgrid is displayed in **Figure 1**. A solar farm, a wind turbine, a lithium-ion battery, a hydrogen fuel cell, and a utility grid are all part of the proposed AC microgrid. The AC microgrid (MG) architecture has been chosen instead of a DC architecture owing to its compatibility with existing

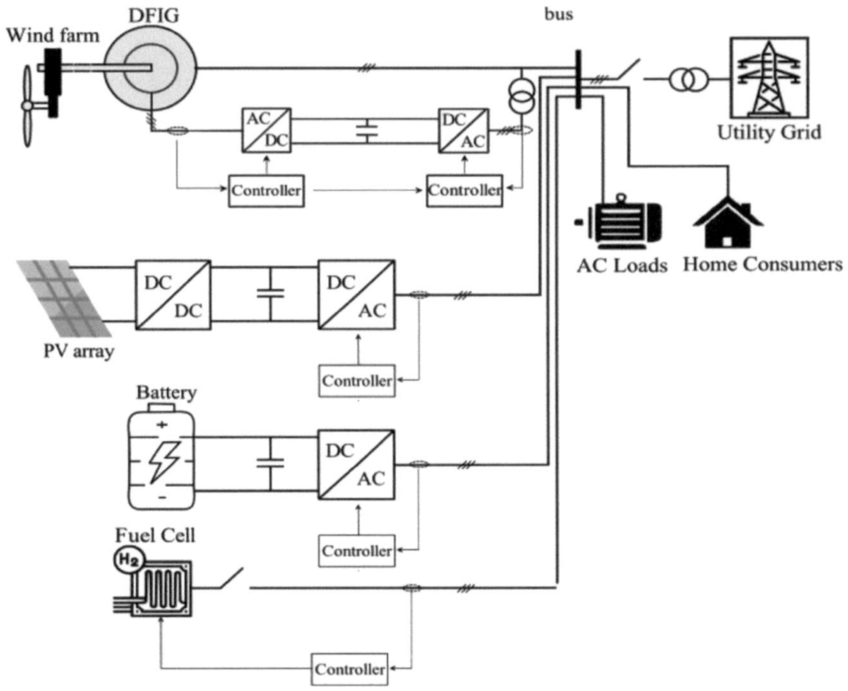

Figure 1.
General structure of the proposed microgrid.

infrastructure and greater flexibility in power distribution network. AC MGs are emerging and becoming more attractive structures with integration of RES based DG units and ESSs in order to manage our future energy demands based flexibility, digitalization and energy transition, but also as a viable and reliable solution to the population without access to energy or with poor energy supply to effectively reduce the greenhouse gas emissions. The microgrid requires a battery to handle electrical loads during periods of low renewable energy generation because renewable energy supply is highly variable and depends on the environmental conditions. The microgrid may be confronted with extended periods of low irradiance and low wind speed, potentially resulting in a fully discharged battery. In this instance, if the utility grid is unavailable, a hydrogen-fueled fuel cell can be employed to meet the load demand.

2.1 PV system

Multiple photovoltaic arrays with a combined capacity of 60 kW make up the solar farm. A boost converter is used to boost the DC output of the photovoltaic arrays and keep the PV modules' generation at its highest level. A three-phase full-bridge inverter is employed because the microgrid's PCC is a three-phase AC system, and the three-phase square waves from the inverter are subsequently filtered using an LCL filter. **Figure 2** illustrates the model [39].

2.1.1 Photovoltaic array

The MATLAB-Simulink special power system library provides the photovoltaic array model/block/subsystem, which we have utilized in the microgrid simulation.

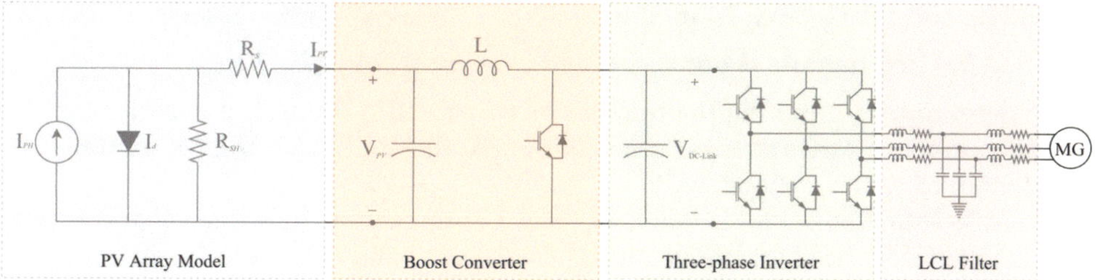

Figure 2.
Model of the PV system [39].

It is a five- parameter single-diode model that uses a light-generated current source (I_{ph}) and a diode current (I_d) to simulate an ideal PV cell with a series resistance (R_s) and a parallel-coupled shunt resistance (R_{sh}) to simulate a more practical solar cell and more accurately describes the solar cells' power losses. The one-diode model is one of the most popular models because of the good compromise between simplicity and precision [7, 24, 33, 37–39].

The I-V characteristic of the solar cell can then be derived by using the single-exponential Shockley equation for the diode, and the resistances to get Eq. (1) [33].

$$I = I_{ph} - I_0\left(e^{\frac{V+R_sI}{nV_t}} - 1\right) - \left(\frac{V + R_sI}{R_{sh}}\right) \quad (1)$$

where I_0 is the reverse saturation current, n is the ideality factor of the diode, and V_t is the thermal voltage. Eq. (2) describes the light-generated current I_{ph} which is based on the value of irradiance (G), the cell temperature (T_c), the STC (Standard Test Condition) of the irradiance (G_{ref}), and cell temperature (T_{ref}), the temperature coefficient k_i ($A/°C$), and the short-circuit current at STC (I_{sc}) [29].

The reverse saturation current is given by Eq. (2), where the I_{sc} is the short-circuit current, V_{oc} is the open-circuit voltage, V_{to} is the STC thermal voltage, E_g is the energy bandgap of the semiconductor, and the energy bandgap at T = 0 K (E_{go}) [29].

$$I_0 = \frac{I_{SC}e^{\left(\frac{E_{go}}{V_{to}} - \frac{E_g}{V_t}\right)}}{e^{\left(\frac{V_{OC}}{nN_SV_{to}}\right)} - 1} \cdot \left(\frac{T_C}{T_{ref}}\right)^3 \quad (2)$$

Next, the semiconductors energy bandgap value at any cell temperature (T_c) is described by Eq. (3), where the α_{gap}, and β_{gap} are the characteristic parameters of the semiconductor [33, 35].

$$E_g = E_{go} - \frac{a_{gap}\, T_C^2}{b_{gap} + T_C} \quad (3)$$

However, several solar cells are connected in series in a photovoltaic module, and some modules may have multiple parallel branches of the series connections. The solar cell equations can be scaled up by representing the number of solar cells connected in series as (N_s) and the number of parallel branches (N_p). The scaling is performed on the module current ($I_m = N_p I$), module voltage ($V_m = N_s V$), module series resistance $\left(R_{sm} = \frac{N_s}{N_p} R_s\right)$ and module shunt resistance $\left(R_{shm} = \frac{N_s}{N_p} R_{sh}\right)$ [29, 32].

Furthermore, by denoting the number of PV modules connected in series by (N_{sm}) and the number of series strings connected in parallel by (N_{pm}), the PV modules can be scaled up to create a PV array.

The MPP (Maximum Power Point) of the PV array current (I_{mg}) and PV array voltage (V_{mg}) can therefore be characterized using the Eqs. (4) and (5), respectively [33].

$$I_{mg} = N_{pm}\left(\frac{I_{mmR}}{1000}G + \left(\frac{dI_{scm}}{dT}\right)(T_c - T_{ref})\right) \tag{4}$$

$$V_{mg} = N_{sm}\left(N_s V_t \ln\left(1 + \frac{Iscm - I_{mg}}{I_{scm}}\left(e^{\frac{V_{ocm}}{N_s V_t}} - 1\right)\right) - I_{mg} R_{sm}\right) \tag{5}$$

Where $\mathbf{I}_{mmg}$ is the rated MPP current of the module at STC, $\mathbf{I}_{scm}$ is the short-circuit current of the module at STC, V_{ocm} is the open-circuit voltage of the module at STC. The PV- module operating temperature (T_c), can then be found for any irradiance condition, and ambient air temperature (T_{air}), as shown in Eq. (6). NOCT is the normal operating cell temperature at an irradiance of 800 W/m^2, and an ambient air temperature of 20°C [33].

$$T_c = T_{air} + \frac{NOCT - 20}{800} G \tag{6}$$

It is worth noting that the single-diode model has poor accuracy for extremely low irradiances, but the two-diode model can be utilized to improve accuracy in these cases. The two-diode model, on the other hand, is substantially slower to simulate because it has seven parameters and two exponential components, and the accuracy at low irradiances has no effect on the overall output power.

2.1.2 Boost converter and MPPT

The boost converter is a DC/DC converter that increases the output voltage through active switching. The boost converter contains an input capacitor, an inductor, an IGBT, a diode, and an output capacitor and is built using blocks from the MATLAB-Simulink special power system library. **Figure 3** shows the MATLAB-Simulink model.

When the gate of the IGBT receives a square wave of sufficient magnitude, it conducts (ON state), creating a short circuit between the inductor and the negative input. The inductor on the input side stores energy in the magnetic field, and the current will only pass through the IGBT because the diode, capacitor, and load all have much greater impedances.

There is no path through the IGBT when it is turned off, and the abrupt drop in current causes the inductor to generate a back EMF with the polarity of the voltage across it during the ON period. As a result, two voltages are generated, one from the supply and the other from the inductor. The current going through the diode is now charging the capacitor and powering the load at the same time. Even if no current passes from the input to the output during the ensuing on-period, the output capacitor will retain charge and continue to power the load [7].

Because the output voltage remains constant in a steady state, the integral of the inductor voltage over one period is zero. Eq. (7) can therefore be used to explain the

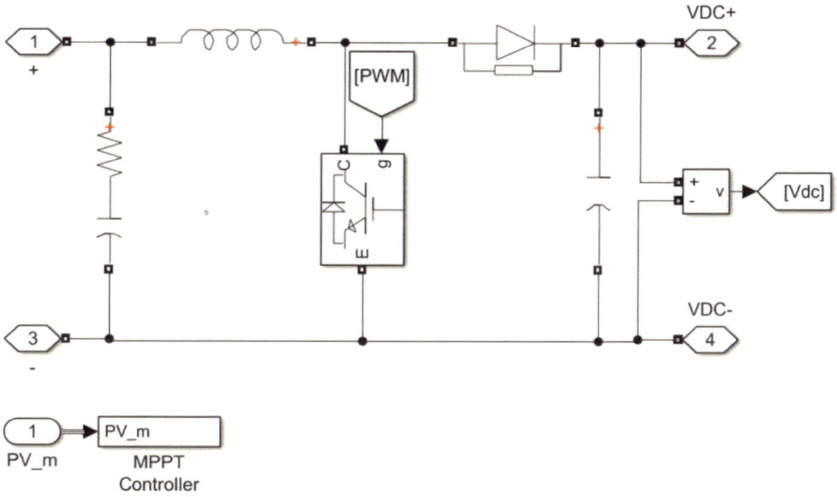

Figure 3.
MATLAB-Simulink model of the boost converter modeled with blocks from Simscape/SPS library and the MPPPT controller based on a MATLAB function.

dynamics in CCM (Continuous Conduction Mode). After that, divide both sides of Eq. (7) with.

T_s and rearrange to get Eq. (8) [7].

$$V_i t_{on} + (V_i - V_o)t_{off} = 0 \tag{7}$$

$$\frac{V_o}{V_i} = \frac{T_s}{t_{off}} = \frac{1}{1-D} \tag{8}$$

Where V_i is the input voltage, V_o is the average output voltage, t_{on}, and t_{off} are the time the IGBT is switched on, and off during one period respectively, T_s is the switching period, and D is the duty cycle.

We can define the minimal amount of inductance required to function at CCM when choosing the inductor. Eq. (10) may be used to compute the critical inductance value, whereas Eq. (9) can be used to calculate the duty cycle [7].

$$D = 1 - \frac{V_{mpp}}{V_{o-nom}} \tag{9}$$

$$L_c = \frac{V_{mpp}D}{\Delta I_L f_{sw}} \tag{10}$$

Where V_{mpp} is the PV-maximum array's rated voltage at maximum irradiance and lowest ambient temperature. The nominal output voltage is V_{oNom}, the inductor ripple current is ΔI_L, and the switching frequency is f_{sw}. Eqs. (11) and (12) may then be used to compute the capacitance required at the input and output [7].

$$C_{in} = \frac{\Delta I_L}{8\Delta V_{pv} f_{sw}} \tag{11}$$

$$C_{out} \geq \frac{V_o D}{f_{sw}\Delta V_o R} \tag{12}$$

2.1.3 MPPT algorithm

To always generate the maximum possible power with the PV array, an MPPT (Maximum Power Point Tracking) algorithm is used [37]. The algorithm that is used in this simulation is called P&O (Perturb & Observe) and its flowchart is visualized in **Figure 4**. The MPPT generates a voltage reference in the model. The duty cycle for the PWM generation is then produced by feeding the difference between the observed voltage and the voltage reference into a PI controller.

2.1.4 Three-phase square-wave inverter

The three-phase full-bridge inverter is a switching transistor-based DC/AC converter. A large-value capacitor is utilized in the DC-link to smooth out the input voltage to the inverter since VSIs (Voltage Source Inverters) depend on a consistent DC source. The square-formed sine wave that the inverter outputs as AC voltage must first be filtered before reaching the PCC [38, 40, 41].

The six IGBTs that make up the three-phase, two-level inverter are split into three at the top that are connected to one of the phase outputs from the positive DC input and three more that are connected to the same output from the negative DC input. To

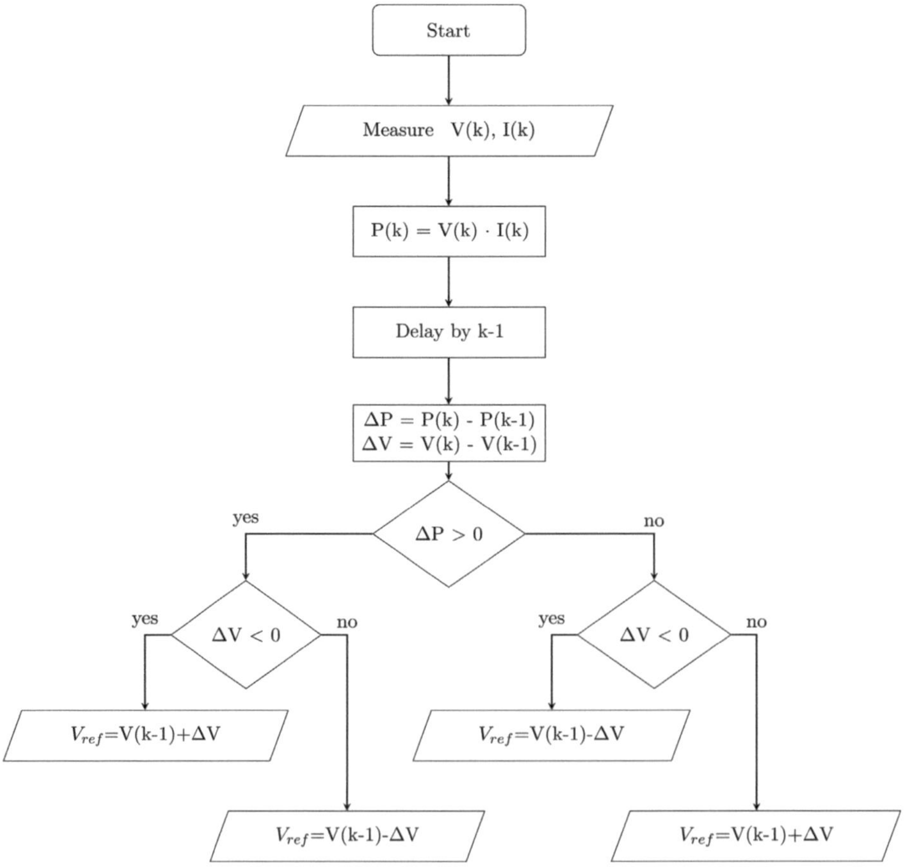

Figure 4.
MATLAB-Simulink MPPT algorithm used for controlling the duty cycle of the boost converter.

prevent a short circuit, it is crucial that never both of an IGBT's top and bottom levels conduct at once. **Figures 5** and **6** show the three-phase square wave inverter's basic setup and the voltage for each phase, respectively.

2.1.5 LCL filter

For applications that employ a VSI, a filter is necessary to improve the performance of the feedback control and reduce harmonics. There are many different filters that can be utilized, but in this instance an LCL filter is used [27, 28, 37, 38]. The LCL filter offers greater attenuation than using a single high-value inductor. Even at power levels of hundreds of kW, the capacitor and inductor values might be minimal. The current ripple, filter size, and switching ripple attenuation must all be considered while creating an LCL filter for a VSI. Additionally, both the inductor and the capacitor may contribute if the controller is used to regulate reactive power, necessitating

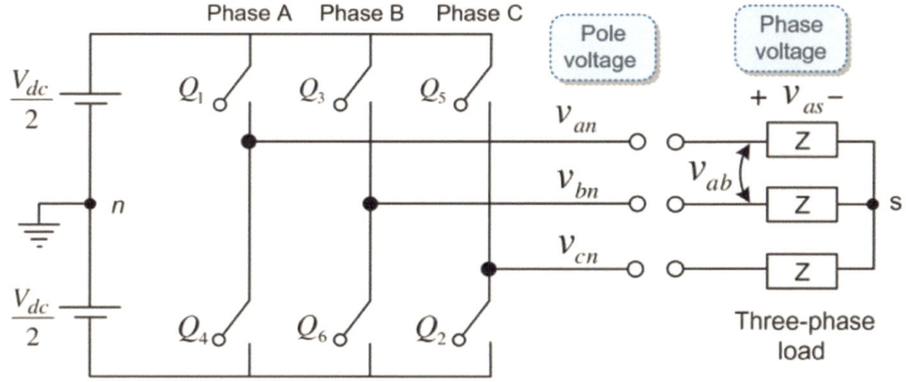

Figure 5.
Basic configuration of the three-phase square wave inverter.

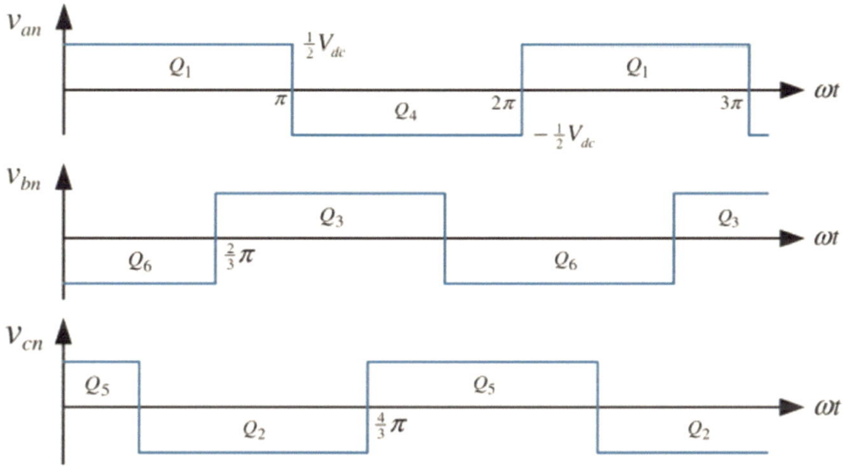

Figure 6.
Three-phase voltages of the three-phase square wave inverter.

damping to prevent resonance The maximum ripple current $\mathbf{I}_{max}$ can be calculated with Eq. (15). In this equation, it is assumed that the maximum peak-to-peak current happens at the inverter modulation factor (m = 0.5). Using Eq. (16), the maximum ripple is set to be 10% of the maximum current [40, 41].

$$I_{\max} = \frac{P_n\sqrt{2}}{3V_{ph}} \tag{13}$$

$$\Delta I_{L\max} = 0.1 I_{\max} \tag{14}$$

With this information, the inverter side inductance L_1, the grid side inductance L_2, and the capacitance C_f can be calculated by using Eqs. (17), (18), and (19). The capacitors can be connected either in Δ or Y configuration. The equations below are for Y connection, while for Δ connection the resulting value from Eq. (19) is divided by 3 and the same goes for the damping resistor g_f [37].

$$L_1 = \frac{V_{dc}}{6f_{sw}\Delta I_{L\max}} \tag{15}$$

$$L_2 = \frac{\sqrt{\frac{1}{k_a^2}+1}}{C_f f_{sw}^2} \tag{16}$$

$$C_f = x \cdot C_b \tag{17}$$

where k_a is the attenuation factor, and x is the maximum power factor variation as seen by the grid. Next, the resonant frequency and damping resistor can be calculated by using Eqs. (20), (21) [40, 41].

$$\omega_{res} = \sqrt{\frac{L_1 + L_2}{L_1 L_2 C_f}} \tag{18}$$

$$R_f = \frac{1}{3\omega_{res} C_f} \tag{19}$$

It is important that the resonant frequency is kept between the limits in Eq. (22) [40, 41].

$$10f_g < f_{res} < 0.5f_{sw} \tag{20}$$

All the parameters that have been used for the model can be seen in **Table 1**.

2.2 Wind farm

The wind farm consists of a synchronous machine, which is driven by a wind turbine coupled with a diode rectifier and a boost converter that is used to increase the DC-link voltage. A full-bridge inverter is then used to convert the DC power back to three-phase AC. After that, an LCL filter is employed to remove harmonics and smooth out the square waves coming from the inverter. The model is drawn from the MATLAB-Simulink Simscape/special power systems package, where the parameters are designed to satisfy the microgrid requirements [42].

f_g	Grid Frequency	50 Hz
f_{sw}	Switching Frequency	10 kHz
P_n	Nominal Power	60 kW
V_g	Phase Grid Voltage	230 V
V_{dc}	DC-Link Voltage	700 V
x	Maximum Power Variation	20%
k_α	Attenuation Factor	20%
L_1	Inverter Side Inductor	1.4 mH
L_2	Grid Side Inductor	0.374 mH
C_f	Capacitor Filter	160 μF
R_f	Damping Resistor	0.4528

Table 1.
Parameter for the LCL filter.

2.2.1 Wind turbine

The wind turbine is modeled using the wind speed V_w, the pitch angle β, and the rotor speed ω_t as input parameters. The equations used for modeling the wind turbine are shown below in Eqs. (23), and the Simulink block model is shown in **Figure 7**. The mechanical system is based on the equation of motion that is displayed in Eq. (24) [43].

$$P_m = \frac{1}{2}\rho A v_w^3 C_p(\lambda, \beta) \tag{21}$$

$$T_{mech} - T_{elec} = I\frac{d\omega}{dt} \tag{22}$$

2.2.2 Synchronous machine

The synchronous machine model has been taken from the specialized power system library and represents the dynamics of the stator, field, and damper windings. It is modeled in the dq-reference frame and is based on Eqs. (25) (33) [6, 11].

$$V_d = -i_d R_s - \omega\psi_q + \frac{d\psi_d}{dt} \tag{23}$$

$$V_q = -i_q R_s - \omega\psi_d + \frac{d\psi_q}{dt} \tag{24}$$

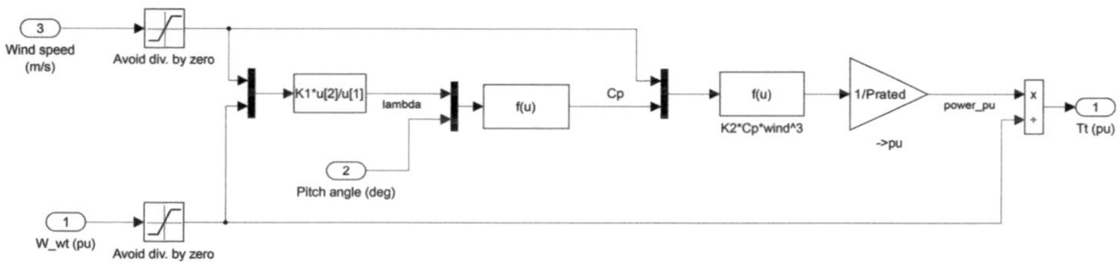

Figure 7.
The MATLAB-Simulink block model of a wind turbine.

$$V_0 = -i_0 R_0 + \frac{d\psi_0}{dt} \tag{25}$$

$$V_{fd} = V_{fd} = \frac{d\psi_{fd}}{dt} + r_{fd} i_{fd} \tag{26}$$

$$\frac{d\psi_{kd}}{dt} + R_{kd} i_{kd} = 0 \tag{27}$$

$$\frac{d\psi_{kq1}}{dt} + R_{kq1} i_{kq1} = 0 \tag{28}$$

$$\frac{d\psi_{kq2}}{dt} + R_{kq2} i_{kq2} = 0 \tag{29}$$

$$\begin{bmatrix} \psi_d \\ \psi_{kd} \\ \psi_{fd} \end{bmatrix} = \begin{bmatrix} L_{md} + L_f & L_{md} & L_{md} \\ L_{md} & L_{lkd} + L_{f1d} + L_{md} & L_{f1d} + L_{md} \\ L_{md} & L_{f1d} + L_{md} & L_{fd} + L_{f1d} + L_{md} \end{bmatrix} \begin{bmatrix} -i_d \\ i_{kq1} \\ i_{kq2} \end{bmatrix} \tag{30}$$

$$\begin{bmatrix} \psi_q \\ \psi_{kq1} \\ \psi_{kq2} \end{bmatrix} = \begin{bmatrix} L_{mq} + L_f & L_{mq} & L_{mq} \\ L_{mq} & L_{mq} + L_{kq1} & L_{mq} \\ L_{mq} & L_{mq} & L_{mq} + L_{kq2} \end{bmatrix} \begin{bmatrix} -i_d \\ i_{kd} \\ i_{fd} \end{bmatrix} \tag{31}$$

All the nomenclatures of the parameters in the equations can be found in [44].

2.2.3 Back-to-Back boost converter

A library from MathWorks' current collection was also used to select the back-to-back boost converter. It is a part of the library's specialized power systems block for wind turbine subsystems. **Figure 8** depicts the model. Three-phase AC from the synchronous machine is fed into the back-to-back boost converters, which are then transformed into DC by a diode bridge (rectifier). The voltage is subsequently increased by the boost converter; for further information on the boost converter, see Section 2.1.1. The three-phase square wave inverter, which is discussed in Section 2.1.3, is then given the stepped-up DC voltage.

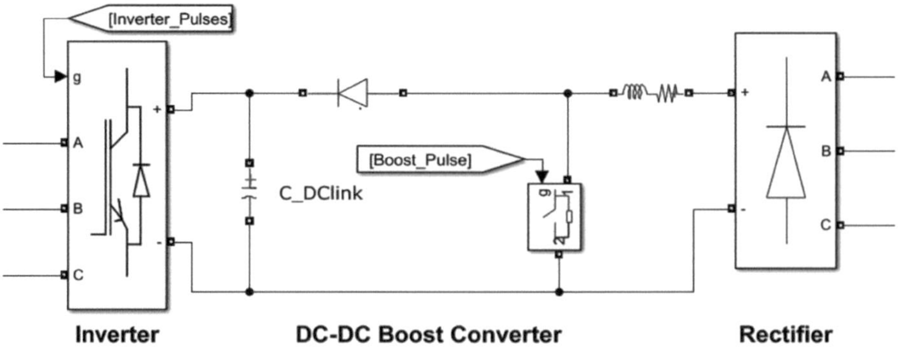

Figure 8.
The MATLAB-Simulink model of the back-to-back converter with DC-DC boost converter in DC-link.

2.3 Energy storage system

It is crucial to have the ability to store energy during periods of high-power generation and use it during periods of low generation since the renewable energy sources in the microgrid are very intermittent and dependent on the environment. Additionally, the ESS (Energy Storage System) can be utilized to peak-shave, trade with the grid, and enhance the microgrid's dependability and power quality. It is made up of an L-filter, a three-phase square wave inverter, and a lithium-ion battery bank as illustrated in **Figure 9**. The Simulink model of the ESS is displayed in **Figure 10**.

2.4 Lithium-ion battery

Eq. (34) describes the discharging process of the lithium-ion battery, while Eq. (35) describes the charging process of the battery [28].

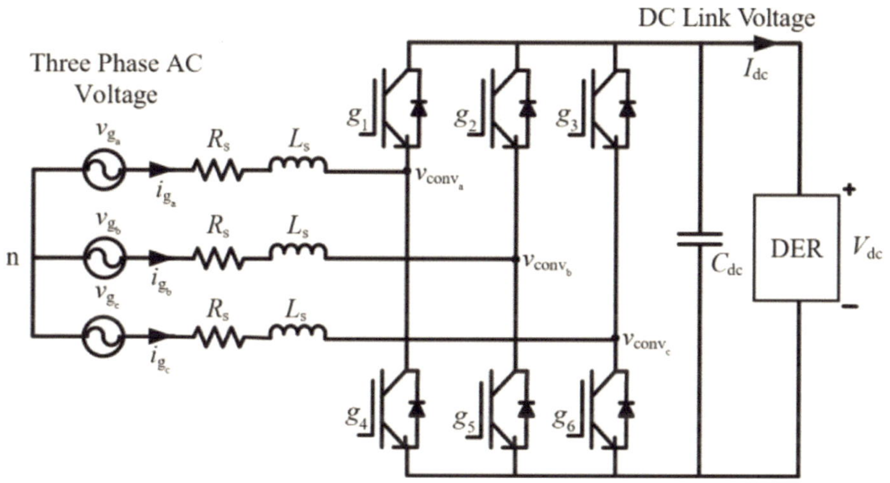

Figure 9.
Circuit diagram of the bidirectional grid converter.

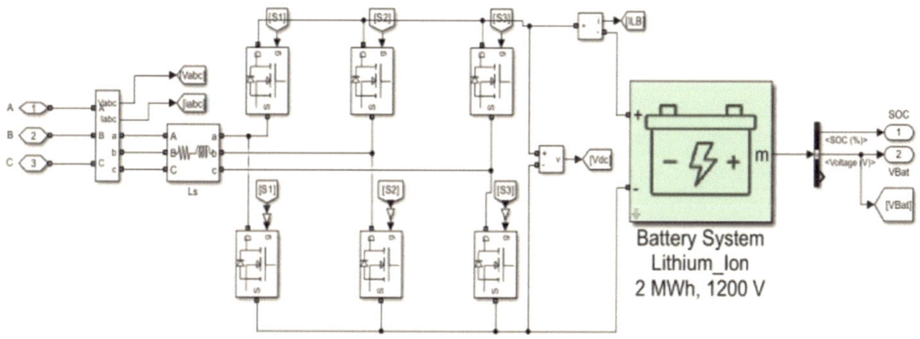

Figure 10.
The MATLAB-Simulink model of the battery storage system.

$$f_1(it, i^*, i) = E_0 - K\frac{Q}{Q - it}i^* - K\frac{Q}{Q - it}it + Ae^{-B.it} \tag{32}$$

$$f_2(it, i^*, i) = E_0 - K\frac{Q}{it + 0.1Q}i^* - K\frac{Q}{Q - it}it + Ae^{-B.it} \tag{33}$$

Where E_0 is the constant voltage, K is the polarization constant (V/Ah), i^* is the low- frequency current dynamics, i is the battery current, it is the extracted capacity in Ah, Q is the maximum battery capacity, A is the exponential voltage, and B is the exponential capacity (**Figure 11**).

2.5 Hydrogen fuel cell

The backup power source is present so that the microgrid can continue to operate in islanded mode even when the energy storage system's state of charge (SOC) is low. The backup power source in this microgrid is a hydrogen fuel cell. The fuel cell is modeled as a dependent voltage source with a series internal resistance and internal diode as displayed in the equivalent electric circuit as depicted in **Figure 12**. The inverter used with the fuel cell is the same as the one used with the battery illustrated in **Figure 9** and it also uses the same control architecture.

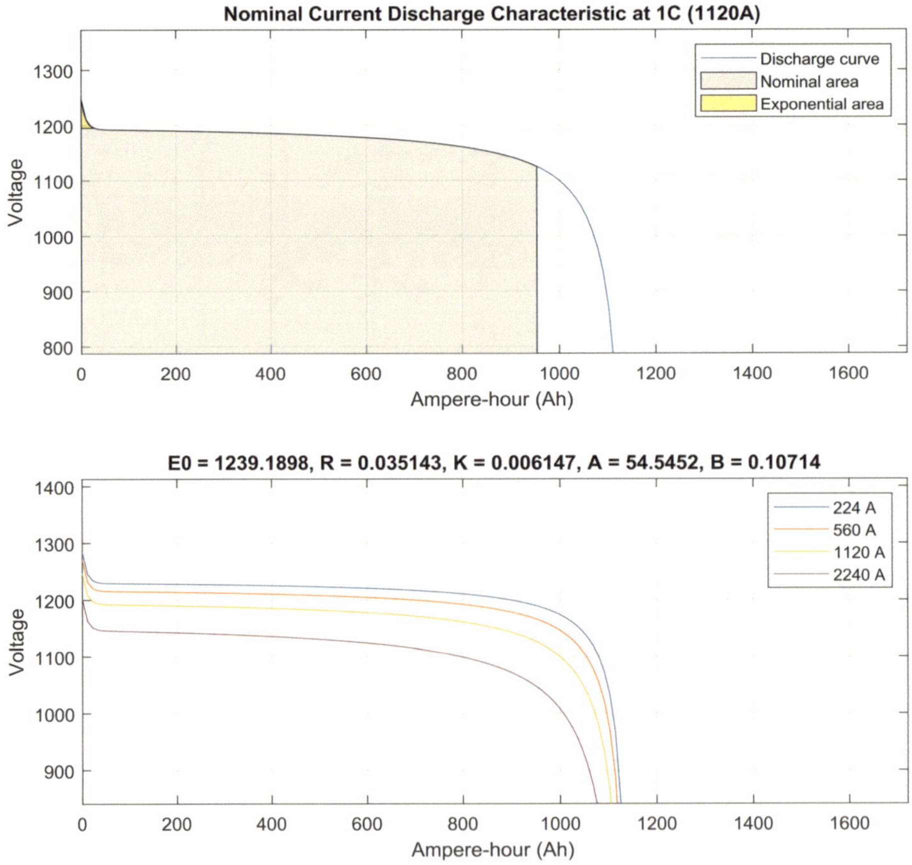

Figure 11.
Discharge characteristics of the battery storage model at different currents.

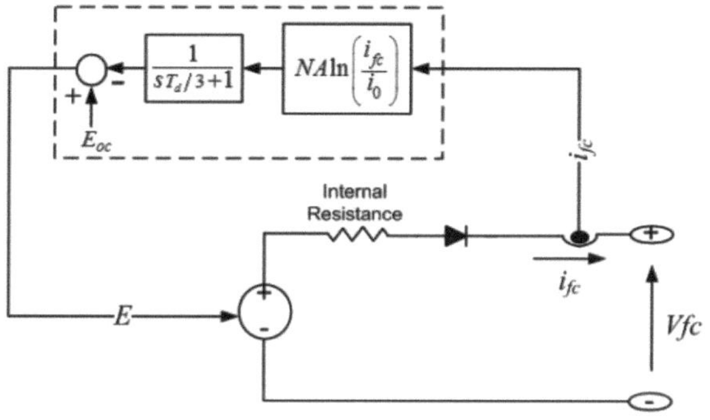

Figure 12.
Equivalent circuit diagram of the fuel cell.

3. Development and testing of an energy management system (EMS) algorithm

3.1 The proposed EMS algorithm

The suggested approach for controlling the energy flow is depicted in a flowchart in **Figure 13**. The suggested EMS additionally considers the battery limitations, which state that the battery should never be charged below or above the limits (SOC), nor should the maximum charging or discharging current, denoted by $P_{bat,max}$, be exceeded. The algorithm was created using a MATLAB function. Following the validation of the SOC restrictions, the algorithm checks to see if the microgrid is generating more energy than the load is using; if so, the battery is charged in accordance with the restrictions. If the generation is lower than the load, the battery must discharge, or if the SOC is low, the fuel cell must be engaged.

3.2 Testing scenarios and results

A number of scenarios were developed to test the effectiveness and dependability of the suggested EMS in the islanded mode. The results are presented in **Figures 14** and **15**, and a summary of these situations is provided in the **Table 2**. Intially, the load reference were set to 10 kW, and the wind and PV reference were increased to generate the full power. This creates an unbalance as the generation is much larger than the load and since the battery can only absorb 50 kW, the generated power had to be limited to by disabling the MPPT and reducing the PV power reference. That is exactly what the EMS did as it can be seen that the MPPT was disabled shortly after 0 and the PV power was limited while making sure that the battery is charging with a maximum power of 50 kW and the load is kept stable at 10 kW. Next, a step in the load active power reference was applied from 10 to 50 kW and at the same instant, a step in the reactive power was also applied from 0 to 44 kVAR, making the load power factor 0.75 lagging. The controller responds appropiately by increasing the PV power reference to meet the load demand while keeping the battery charging at 50 kW. The reactive power demand is also met by the battery controller. After that, a step in the load reactive power was applied from 44 to −44 kVAR which changes the power factor from lagging to leading and the battery also was able to absorb the reactive

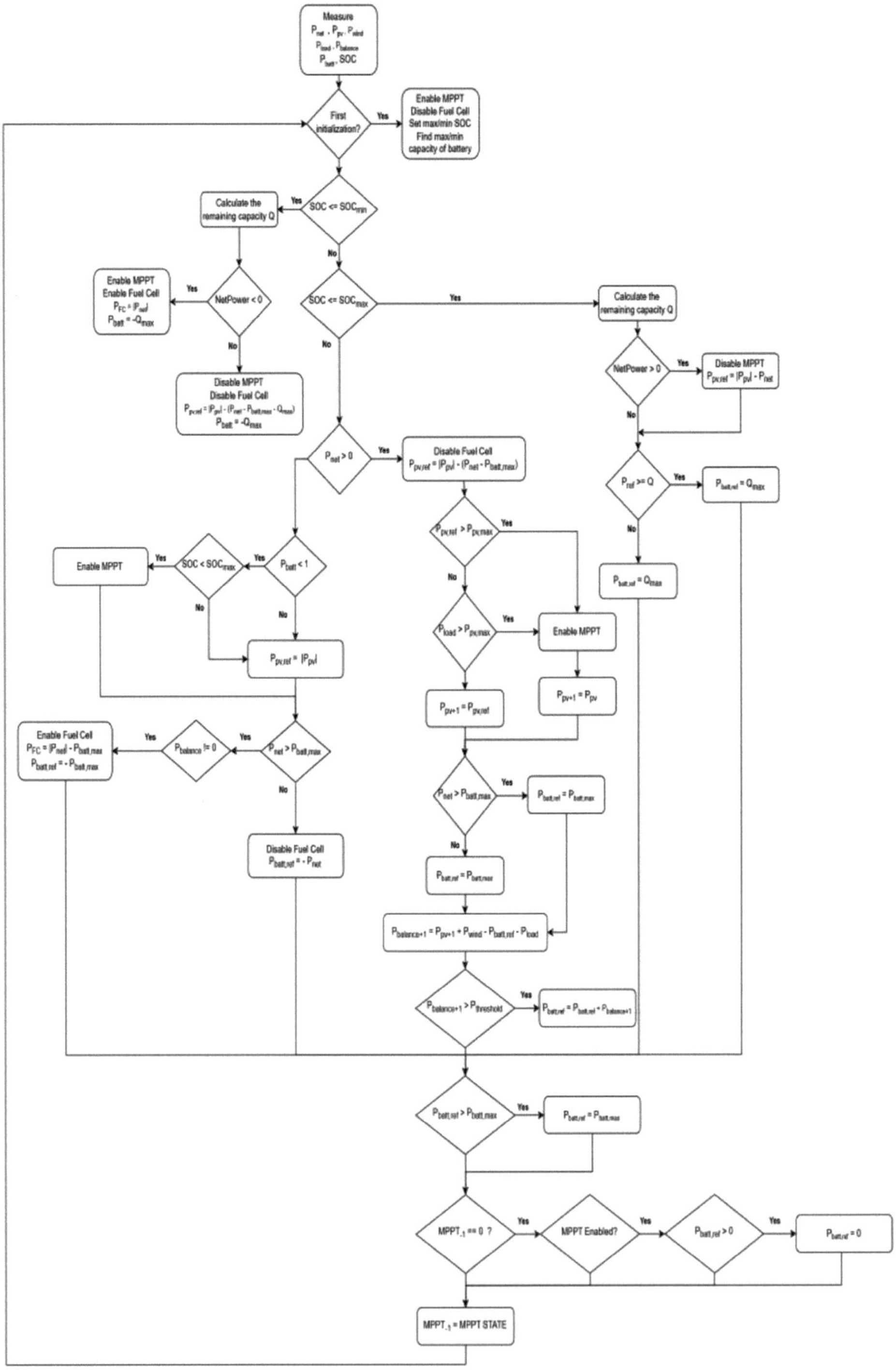

Figure 13.
Flow-chart block diagram of the EMS algorithm.

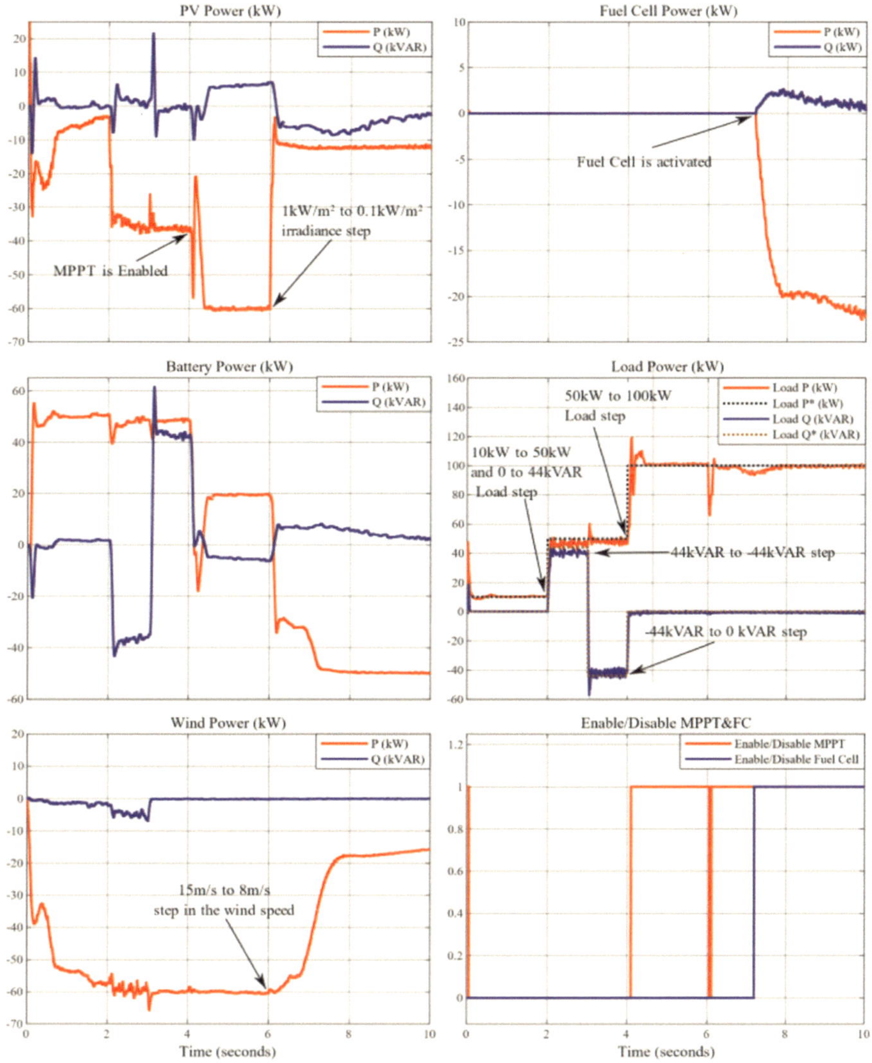

Figure 14.
Active and reactive power of different components in the microgrid under loads steps.

power without any issues. At 4 seconds, the reactive power was reset to 0 and another step in the active power was applied from 50 to 100 kW to test the system at full load. Since the load is now much higher, the EMS enabled the MPPT to ensure taking the maximum power of 60 kW available from the PV. The remaining power came from the wind and any excess power was used to charge the battery. Finally, at 6 seconds, the generation was reduced evem further to test the scenario where the load is higher than the generation. A step in the irradiance from 1000 to 100 W/m^2 was applied to the PV and a similar step in the wind speed was applied from 15 to 8 m/s. As the wind power generation was gradually reducing, the EMS sent control commands to the battery to supply the remaining power and the battery started discharging up to

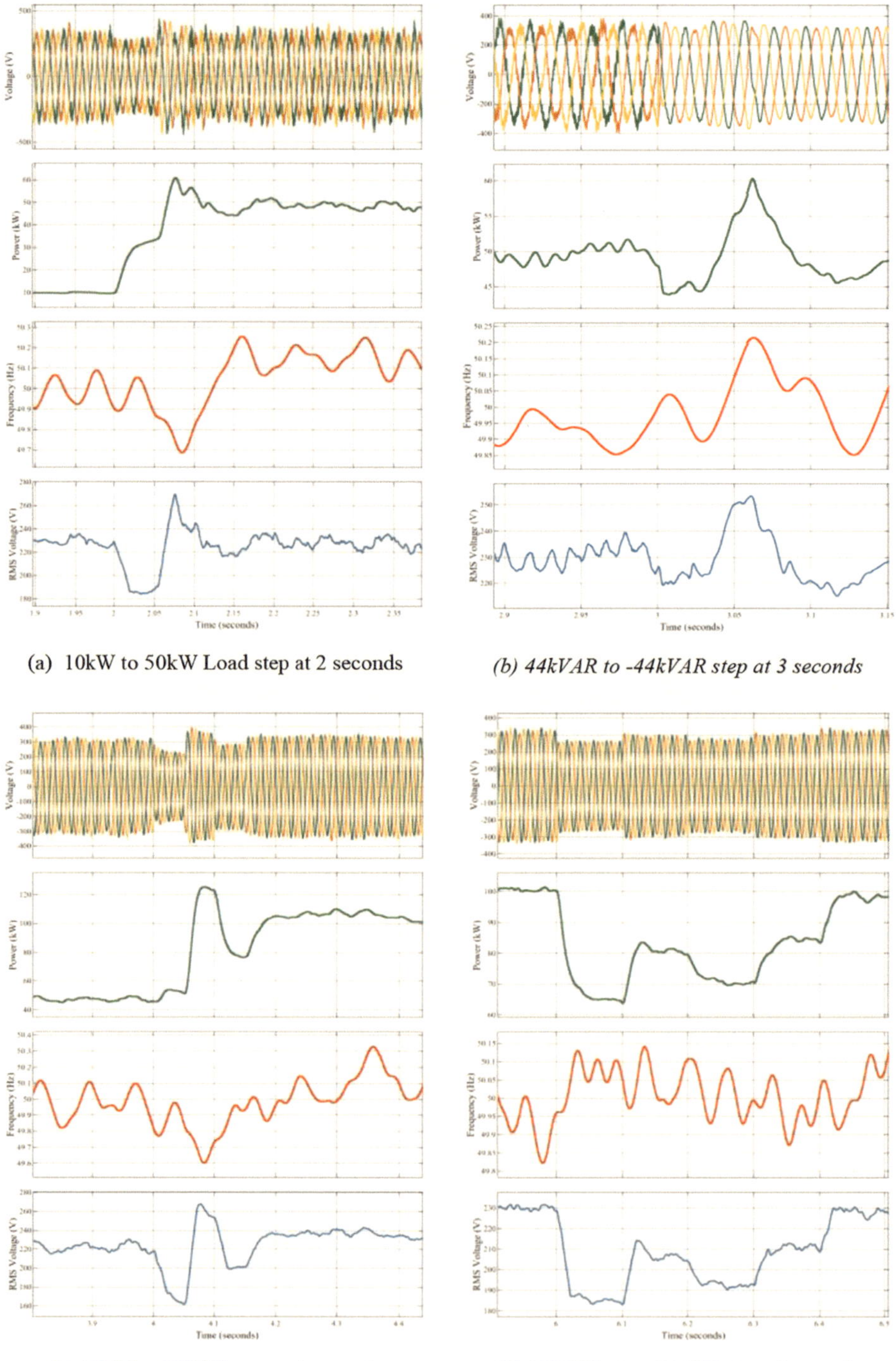

(a) 10kW to 50kW Load step at 2 seconds

(b) 44kVAR to -44kVAR step at 3 seconds

(c) 50kW to 100kW step at 4 seconds

(d) 1000W/m2 to 100W/m2 at 6 seconds

Figure 15.
Three-phase voltage, power, frequency, and RMS value of the load under different scenarios. (a) 10 to 50 kW load step at 2 seconds. (b) 44 to −44 kVAR step at 3 seconds. (c) 50 to 100 kW step at 4 seconds. (d) 1000 to 100 W/m2 at 6 seconds.

Scenario Time	Description
At 2 seconds	Step change in the active power reference from 10 to 50 kW and in the reactive power from 0 to 44 kVAR
At 3 seconds	Step change in the reactive power reference from 44 to – 44 kVAR
At 4 seconds	Step change in the active power reference from 50 to 100 kW and in
At 6 seconds	Step change in the irradiance from 1000 to 100 W/m2 and in the

Table 2.
Summary of the scenarios tested with energy management system in islanded.

−50 kW, once the battery reached its maximum discharging power, the fuel cell had to be enabled to supply the remaining load power and that is what happened at around 7 seconds. From there on, all the microgrid sources were working in tandem to keep the load power at 100 kW while the EMS adjusted the fuel cell reference based on the power generated by the wind and PV system. These scenarios clearly illustrate that the proposed energy management system is robust and can succesfully control the microgrid under various conditions.

The wind and solar references were initially maximized to produce the maximum electricity with the load reference set at 10 kW. Due to the imbalance caused by the generation being significantly greater than the load and the battery's ability to store only 50 kW of power, the generated power has to be constrained by turning off the MPPT and lowering the PV power reference. The MPPT was disabled shortly after zero, and the PV power was limited to ensure that the battery was charging with a maximum power of 50 kW and the load was maintained at 10 kW. This is exactly what the EMS performed.

The load's active power reference was then increased from 10 to 50 kW, and at the same time, the load's reactive power was increased from 0 to 44 kVAR, resulting in a load power factor of 0.75 lagging. In order to fulfill the load requirement, the controller increases the PV power reference as necessary, keeping the battery charging at 50 kW. The battery controller also satisfies the demand for reactive power. The power factor was then changed from lagging to leading by applying a step in the reactive power of the load from 44 to −44 kVAR, and the battery was able to absorb the reactive power without any problems.

To test the system at maximum load, the reactive power was reset to 0 at 4 seconds and another step in the active power was applied from 50 to 100 kW. Due to the increased load, the MPPT was enabled to use the full 60 kW of available PV power thanks to the EMS. Wind provided the remaining energy, and any extra was used to recharge the batteries. To test the condition where the demand is more than the generation, the generation was further decreased at 6 seconds. The PV received a step-down in irradiance from 1000 to 100 W/m^2 and a comparable step-down in wind speed from 15 to 8 m/s.

The fuel cell had to be enabled in order to supply the remaining load power, which occurred at roughly 7 seconds as the wind power generation rapidly decreased. The EMS had issued control commands to the battery to deliver the remaining power, and the battery began depleting up to −50 kW. The EMS controlled the fuel cell reference depending on the electricity produced by the wind and PV systems, all the microgrid sources continued to cooperate to maintain the load power at 100 kW. These examples unmistakably show how reliable the energy management system is and how successfully it can operate the microgrid under diverse circumstances.

Following that, the battery's SOC constraints were evaluated using the scenarios depicted in **Figure 16**. **Figure 17(a)** shows a zoomed-in plot of the voltage, frequency,

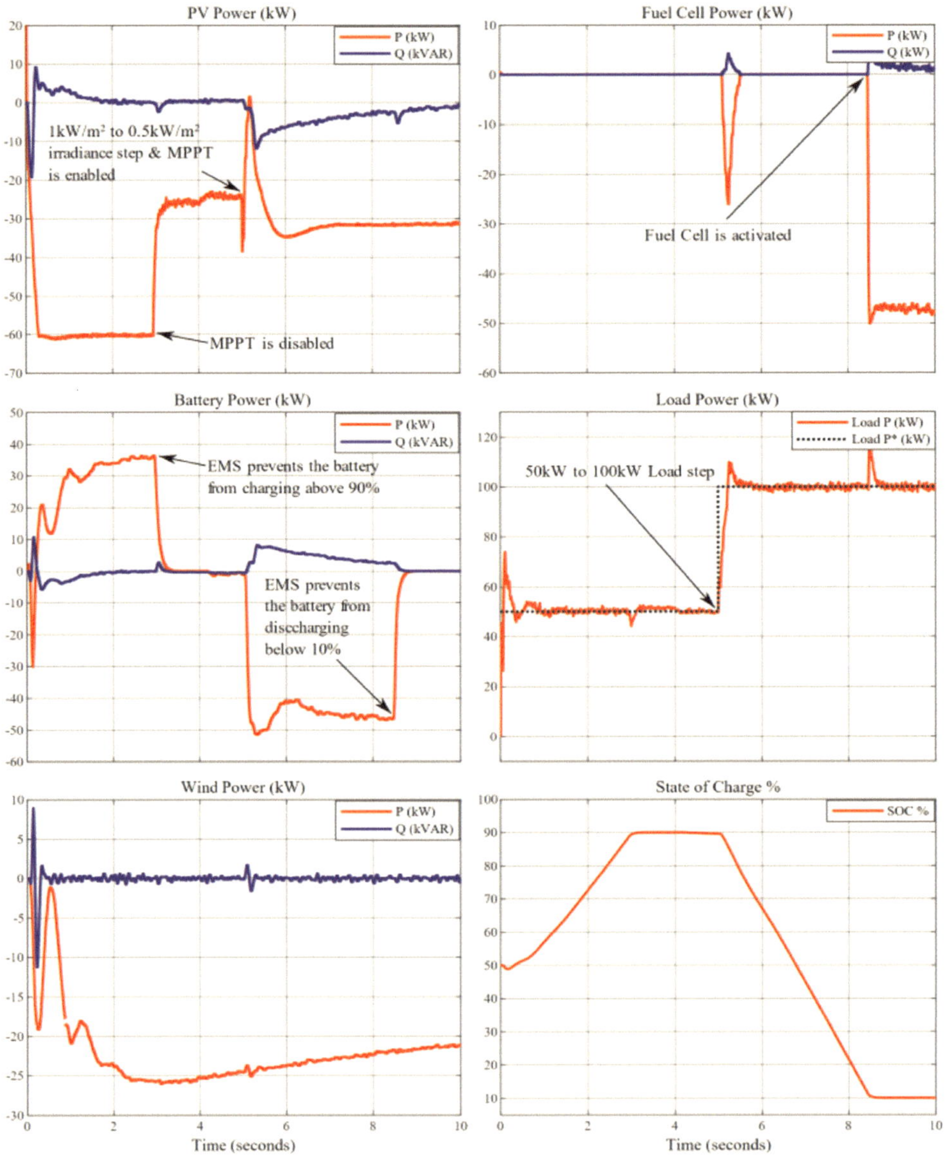

Figure 16.
Active and reactive power of different components in the microgrid to test SOC limiting to protect the battery.

and load consumption at 3 seconds when the controller determined the SOC to be above the limits at 3 s, the battery immediately stopped charging, and the PV could no longer operate at the maximum power point and was given a corresponding reference so the power balance in the microgrid was met. The load is increased to 100 kW at 5 seconds, and the PV irradiance is decreased to 500 W/m^2. To accommodate the significant increase in load demand, this should cause the battery to begin discharging.

After 8 seconds, the battery starts to deplete until it hits 10%, at which point the EMS sends a command to stop discharging the battery. When this occurs, the EMS responds by igniting the fuel cell to start supplying power right away. In

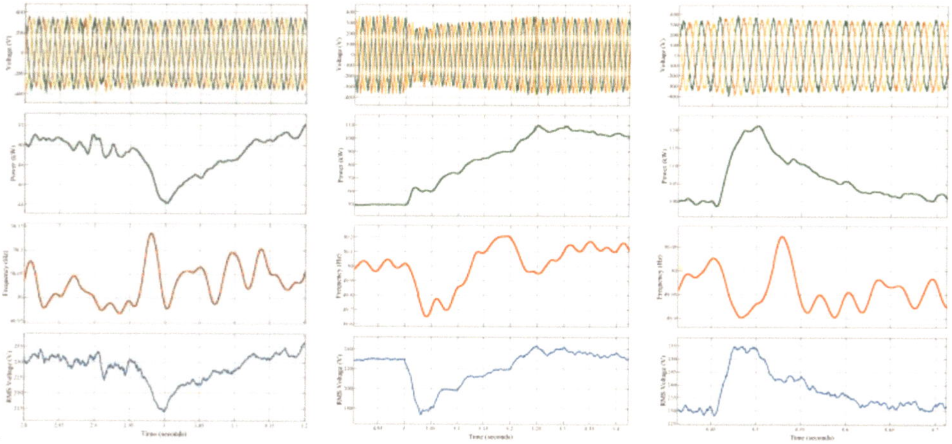

(a) SOC saturation at 90% (b) 50kW to 100kW step at seconds (c) SOC saturation at 10%

Figure 17.
Three-phase voltage, power, frequency, and RMS value of the load under different scenarios. (a) SOC saturation at 90%. (b) 50 to 100 kW step at seconds. (c) SOC saturation at 10%.

Figure 17(c) the voltage, frequency, and load consumption charts are zoomed in. This demonstrates how the EMS can safeguard the battery from overcharging and over-discharging, demonstrating how it accomplishes some of the goals of a BMS (Battery Management System).

4. Conclusion

In summary, this chapter explored the modeling and simulation of microgrid components, including PV system, wind turbine system, battery storage and fuel cell system. The control architecture, developed for the primary control of these components, was based on model predictive control. An energy management system algorithm was successfully designed and developed to control the power flow and to ensure continuous power delivery to the load under all circumstances. The algorithm was tested on a variety of scenarios and proved its robustness and flexibility. When the generation exceeds the load demand and the battery cannot absorb all the excess power, the EMS would disable the MPPT to limit the generation from the solar cell to protect the battery. If the load demand exceeds the generation, the EMS uses the battery to make up for the difference and if the battery power is not sufficient, the fuel cell is activated to provide the rest of the power. Moreover, SOC- based protection scheme was also implemented to ensure that the battery state of charge remains within acceptable limits which increase the lifetime of the battery.

5. Future work

The simulation of the EMS algorithm presented in this paper provides a promising proof- of-concept for its effectiveness in managing electrical energy in a microgrid. However, there are several avenues for further exploration and improvement. One important step for the validation of the EMS algorithm would be to test it on real hardware, such as a Hardware-in-the-Loop (HIL) platform. This would enable us to evaluate the algorithm's performance in a realistic setting, which includes the various noise and uncertainties that can arise in the physical world. Moreover, we could measure the real-time performance of the algorithm and compare it with the simulation results.

Another potential area for future work is to add new equipment to the microgrid and reconfigure the algorithm accordingly. The EMS algorithm was designed to work with a specific set of components, and its performance may be affected by the addition or removal of equipment. Hence, future expansion of the microgrid may require a readjustment of the algorithm to ensure its optimal operation.

Finally, the implementation of the EMS algorithm presented in this paper was focused on a single microgrid. However, in practice, multiple microgrids can be interconnected to form a larger network, and the EMS algorithm must be adapted to this scenario. Future work could explore the development of a hierarchical control scheme that manages multiple microgrids simultaneously.

Funding

This research was partial funded by the EEA and Norway grant/project DOITSMARTER, contract no 2022/337335.

Conflict of interest

The authors declare no conflict of interest.

Author details

Andreas Pedersen, Ibrahim Ahmed and Lucian Mihet-Popa*
Faculty of Information Technology, Engineering and Economics, Østfold University College, Fredrikstad, Norway

*Address all correspondence to: lucian.mihet@hiof.no

References

[1] Chandorkar MC, Divan DM, Adapa R. Control of parallel connected inverters in standalone AC supply systems. IEEE Transactions on Industry Applications. 1993;**29**(1):136-143

[2] IEEE-SA Standards Board. IEEE Standard for Interconnecting Distributed Resources with Electric Power Systems. New York, United States: IEEE; 2003. pp. 1-28

[3] Nicholas JJS-G, Miller W, Price WW. Dynamic modeling of GE 1.5 and 3.6 wind turbine-generators. In: 2003 IEEE Power Engineering Society General Meeting (IEEE Cat. No.03CH37491). Vol. 3. Toronto, ON, Canada: IEEE; 2003. pp. 1977-1983. DOI: 10.1109/PES.2003.1267470

[4] Blaabjerg F, Teodorescu R, Liserre M, Timbus AV. Overview of control and grid synchronization for distributed power generation systems. IEEE Transactions on Industrial Electronics. 2006;**53**(5):1398-1409

[5] Lopes JAP, Moreira CL, Madureira AG. Defining control strategies for microgrids islanded operation. IEEE Transactions on Power Apparatus and Systems. 2006;**21**(2):916-924

[6] Kundur P. Power System Stability & Control. 2nd ed. New York, United States: McGraw Hill; 2022

[7] Hasaneen BM, Mohammed AAE. Design and simulation of DC/DC boost converter. In: 2008 12th Int. Middle East Power Syst. Conf. MEPCON. Vol: 2008. Aswan, Egypt: IEEE; 2008. pp. 335-340

[8] Guerrero JM, Vasquez JC, Matas J, De Vicuña LG, Castilla M. Hierarchical control of droop-controlled AC and DC microgrids—A general approach toward standardization. IEEE Transactions on Industrial Electronics. 2010;**58**(1):158-172

[9] Cho C, Jeon J-H, Kim J-Y, Kwon S, Park K, Kim S. Active synchronizing control of a microgrid. IEEE Transactions on Power Electronics. 2011;**26**(12):3707-3719

[10] Cortes P, Rodriguez J, Silva C, Flores A. Delay compensation in model predictive current control of a three-phase inverter. IEEE Transactions on Industrial Electronics. 2011;**59**(2):1323-1325

[11] Lidula NWA, Rajapakse AD. Microgrids research: A review of experimental microgrids and test systems. Renewable and Sustainable Energy Reviews. 2011;**15**(1):186-202

[12] Marnay C, DeForest N, Lai J. A green prison: The Santa Rita jail campus microgrid. In: 2012 IEEE Power and Energy Society General Meeting. San Diego, CA, USA: IEEE; 2012. pp. 1-2

[13] Rocabert J, Luna A, Blaabjerg F, Rodriguez P. Control of power converters in AC microgrids. IEEE Transactions on Power Electronics. 2012;**27**(11):4734-4749

[14] Yang H, Li Q, Chen W. Microgrid communication system and its application in hierarchical control. In: Smart Power Distribution Systems: Control, Communication, and Optimization. Basel, Switzerland: MDPI, Academic Press; 2019. pp. 179-204

[15] Vasquez JC, Guerrero JM, Savaghebi M, Eloy-Garcia J, Teodorescu R. Modeling, analysis, and design of stationary-reference-frame

droop-controlled parallel three-phase voltage source inverters. IEEE Transactions on Industrial Electronics. 2012;**60**(4):1271-1280

[16] Hirose K, Reilly J, Irie H. The Sendai microgrid operational experience in the aftermath of the tohoku earthquake: A case study. New Energy & Industrial Technology Development Organization. 2013;**308**:1-6

[17] Hu J, Zhu J, Dorrell DG. Generation, and undefined 2014. "Model Predictive Control of Inverters for both Islanded and Grid-Connected Operations in Renewable Power Generations". Vol. 8 (3). New York, United States: Wiley Online Library; 2013. pp. 240–248

[18] Fotopoulou MC, Drosatos P, Petridis S, Rakopoulos D, Stergiopoulos F, Nikolopoulos N. Model predictive control for the energy Management in a District of buildings equipped with building integrated photovoltaic systems and batteries. Energies. 2021;**14**(12): 3369

[19] Dehghanzadeh A, Farahani G, Vahedi H, Al-Haddad K. Model predictive control design for DC-DC converters applied to a photovoltaic system. Electrical Power and Energy Systems. 2018;**103**:537-544

[20] Metry M, Balog RS. An adaptive model predictive controller for current Sensorless MPPT in PV systems. IEEE Open Journal of Power Electronics. 2020;**1**:445-455. DOI: 10.1109/ OJPEL.2020.3026775

[21] Zhao Y, An A, Xu Y, Wang Q, Wang M. Model predictive control of grid-connected PV power generation system considering optimal MPPT control of PV modules. Protection and Control of Modern Power Systems. 2021;**6**:32. DOI: 10.1186/s41601-021-00210-1

[22] Lashab A, Sera D, Guerrero J, Mathe L, Bouzid A. Discrete model predictive control-based maximum power point tracking for PV systems: Overview and evaluation. IEEE Transactions on Power Electronics, Institute of Electrical and Electronics Engineers. 2018;**33**(8):7273-7287

[23] Minchala LI, Zhang Y, Probst O. Design of an Energy Management System for secure integration of renewable energy sources into microgrids. In: Cao W-P, Yang J, editors. Development and Integration of Microgrids. Rijeka: IntechOpen; 2017

[24] Justo JJ, Mwasilu F, Lee J, Jung J-W. AC-microgrids versus {DC}-microgrids with distributed energy resources: A review. Renewable and Sustainable Energy Reviews. 2013;**24**:387-405

[25] Mnih V et al. Playing atari with deep reinforcement learning. arXiv Prepr. 2013. pp. 1-9

[26] Rodriguez J et al. State of the art of finite control set model predictive control in power electronics. IEEE Transactions on Industrial Informatics. 2013;**9**(2):1003-1016

[27] Shafiee Q, Guerrero JM, Vasquez JC. Distributed secondary control for islanded microgrids—A novel approach. IEEE Transactions on Power Electronics. 2013;**29**(2):1018-1031

[28] Zhu C, Li X, Song L, Xiang L. Development of a theoretically based thermal model for lithium ion battery pack. Journal of Power Sources. 2013; **223**:155-164

[29] Hu J, Zhu J, Dorrell DG. Model predictive control of grid- connected

inverters for PV systems with flexible power regulation and switching frequency reduction. IEEE Transactions on Industry Applications. 2014;**51**(1): 587-594

[30] Olivares DE et al. Trends in microgrid control. IEEE Transactions on Smart Grid. 2014;**5**(4):1905-1919

[31] Ahumada C, Cárdenas R, Saez D, Guerrero JM. Secondary control strategies for frequency restoration in islanded microgrids with consideration of communication delays. IEEE Transactions on Smart Grid. 2015;**7**(3): 1430-1441

[32] Xue C, Zhou D, Li Y. Hybrid model predictive current and voltage control for LCL-filtered grid-connected inverter. IEEE Journal of Emerging and Selected Topics in Power Electronics. 2021;**9**(5): 5747-5760

[33] Silvestre S. Chapter 7 - strategies for fault detection and diagnosis of PV systems. In: Yahyaoui I, editor. Advances in Renewable Energies and Power Technologies. Amsterdam, Netherlands: Elsevier; 2018. pp. 231-255

[34] Ram JP, Manghani H, Pillai DS, Babu TS, Miyatake M, Rajasekar N. Analysis on solar PV emulators: A review. Renewable and Sustainable Energy Reviews. 2018;**81**:149-160

[35] Bensalem S, Chegaar M, Herguth A. Band gap dependence with temperature of semiconductors from solar cells electrical parameters. Current Applied Physics. 2017;**17**(1):55-59

[36] Ghani F, Duke M, Carson J. Numerical calculation of series and shunt resistance of a photovoltaic cell using the Lambert W-function: Experimental evaluation. Solar Energy. 2013;**87**(1):246-253

[37] Chee WT, Green TC, Carlos AHA. Analysis of perturb and observe maximum power point tracking algorithm for photovoltaic applications. In: PECon 2008 - 2008 IEEE 2nd Int. Power Energy Conf. Johor Bahru, Malaysia: IEEE; 2008. pp. 237-242

[38] Lee BK, Ehsani M. A simplified functional simulation model for three-phase voltage-source inverter using switching function concept. IEEE Transactions on Industrial Electronics. 2001;**48**(2):309-321

[39] Liserre M, Teodorescu R, Rodriguez P. Photovoltaic inverter structures. In: Grid Converters for Photovoltaic and Wind Power Systems. Chichester, United Kingdom: John Wiley and Sons; 2011. pp. 26-29

[40] Beres RN, Wang X, Liserre M, Blaabjerg F, Bak CL. A review of passive power filters for three-phase grid-connected voltage-source converters. IEEE Journal of Emerging and Selected Topics in Power Electronics. 2016;**4**(1): 54-69

[41] Reznik A, Simões MG, Al-Durra A, Muyeen SM. LCL filter design and performance analysis for grid-interconnected systems. IEEE Transactions on Industry Applications. 2014;**50**(2):1225-1232

[42] Ekanayake JB, Holdsworth L, Jenkins N. Comparison of 5th order and 3rd order machine models for doubly fed induction generator (DFIG) wind turbines. Electric Power Systems Research. 2003;**67**(3):207-215

[43] Teodorescu R, Liserre M, Rodríguez P. Grid converters for photovoltaic and wind power systems. Grid Converters for Photovoltaic and Wind Power Systems. 2010. pp. 125-144

[44] Moeini A, Kamwa I, Brunelle P, Sybille G. Synchronous machine stability model, an update to IEEE Std 1110-2002 data translation technique. In: 2018 IEEE Power Energy Society General Meeting (PESGM). Portland, OR, USA: IEEE; 2018. pp. 1-5

Section 2

Smart Distribution Systems: Methodologies, Real-time Platforms and Testing Methods

Chapter 3

Network Reconfiguration and Reactive Power Compensation Dispatch in Smart Distribution Systems

Ulises Tovar Ramírez, José Horacio Tovar Hernández and Guillermo Gutiérrez Alcaraz

Abstract

A significant challenge is to design strategies to minimize electrical losses in smart distribution systems while observing voltage and feeder loading constraints. Unfortunately, few studies have solved the problem of simultaneously coordinating already installed capacitors banks with network reconfiguration problems. This book chapter presents two methodologies for solving the reconfiguration and reactive power compensation dispatch. Both methodologies are formulated as two-stage solve reconfiguration and reactive power compensation problems with the characteristic of having acceptable computational efficiency and loss reduction close to the optimal solution. In the first stage, network reconfiguration is carried out to discriminate radial configurations that do not satisfy voltage and overload restrictions. In the second stage, a reactive power compensation dispatch is applied to each feasible network configuration by connecting capacitor blocks successively until all available reactive capacity has been used or until a specified loss tolerance has been reached. Finally, switching each capacitor block is carried out using voltage linear sensitivities related to shunt reactive compensation to make a relatively low computational work during the process.

Keywords: distribution systems, capacitor switching, network reconfiguration, reactive power compensation, volt/VAr control

1. Introduction

During the last decades, the great growth in technologies of computers, microprocessors, telecommunications, incorporation of distributed generation (DG) and electronic equipment as ac/dc converters and flexible ac transmission systems (FACTS) at the distribution level has led to opportunities for an advanced supervising and control of electric energy delivery systems (distribution systems), particularly in distribution network automatization, but at the same time, it presents new challenges for accomplishing with the main objective of improving the automatization, control, and efficiency of real-life distribution systems, converging into the infrastructure named smart distribution center. Therefore, almost all the efforts have comprehensively investigated the optimal distribution reconfiguration problem [1].

Distribution systems mainly consist of transformers, several feeders composed of line sections, switches, diverse DG systems, and consumer loads.

In existing distribution systems, global loss minimization is a common objective function because of its high impact on their economic efficiency, and it may be basically carried out in two stages, which are as follows—(i) through feeder reconfiguration by opening or closing switching devices and (ii) reactive power compensation by capacitor bank commutations.

1.1 Distribution system reconfiguration

Distribution system reconfiguration is useful for either planning or real-time control to change the electrical conditions of primary supply feeders to reach, in some sense, an optimal operation point [2]. Reconfiguration is a process for modifying the topological structure of distribution feeders by changing the open/close status of sectional switches and feeder sections to find the minimum loss topology, maintaining voltages between their low and high limits simultaneously, and its radial structure, that is, there is only one path between two points in the same feeder [3]. Sectional switches and feeder sections will be named as switches throughout this chapter.

Many feeders are interconnected to the distribution system keeping a radial structure, which is maintained by properly controlling the status of the associated switches, which are classified as normally open or closed. These switches operate under a feeder fault condition to change their statuses to isolate the faulted feeder section and enable load transferences between adjacent feeders to redistribute current flows without any restriction violation [4]. A smart distribution center supports this process mainly with supervisory and control infrastructure, which helps raise the reliability and efficiency of electricity supplied to the final consumer [5].

Almost all feeders have a mix of industrial, commercial, and residential loads, which, through 24 h, show diverse load variations so that their peak load values occur at different moments. In this sense, reconfiguration permits to transfer of loads to a relatively lesser loaded feeder with the benefit of better voltage regulation and lower electrical losses; also, at the same, it increments security margins and quality of energy supplied [2]. Furthermore, under emergency conditions due to a short circuit in some distribution system points, an important goal is to minimize the close/open switching operations to reduce the load-restoration period. In both situations, the distribution system topology should remain radial [6].

From the above discussion, it is obvious that loss minimization implicitly includes operative cost reductions in the distribution system [7]. Hence, almost of reconfiguration methods have electric loss reduction as the main goal. Furthermore, distribution system reconfiguration is a combinatorial problem involving many open/close switch operations for real-life distribution systems [8]. In fact, a small 33-bus distribution system, and considering that each feeder section has a switch in its extremes, the open/close commutations are 435,897. Then, the development of reconfiguration methods must contain the next features [2]:

- Capacity for estimating loss changes resulting from the reconfiguration process, involving minimal computational work.

- There must be a useful criterion to avoid irrelevant switching actions to reduce the searching space and increment the problem solution efficiency.

Researchers have studied the distribution system reconfiguration problem during the last five decades, developing and using different solution methods. Some of the most reported are the next kind of approaches [9]:

- Heuristic methods
- Metaheuristic algorithms
- Conventional mathematical optimization algorithms
- Hybrid methods involving any of the three above

Heuristic methods (HM) are the most attractive because of their relative simplicity and suitability for operating in real-time environments; however, they do not always obtain the optimal global reconfiguration. This drawback is overridden by metaheuristic methods (MM), but they are more complicated in formulation with larger execution time requirements than HM. Therefore, many MM have been developed using ideas of nature behavior [9], which could be based on genetic algorithms [2], particle swarm optimization [3, 10, 11], tabu search [12, 13], simulated annealing [13–15], variable scaling hybrid differential algorithm [16], ant colony [17, 18], plant growth simulation [19, 20], bacterial foraging [21], gray wolf [22], salp swarm [23], symbiotic organism search, hybrid cuckoo search [24], harmony search [25], and binary gravitational search [26], among others. On the other hand, mathematical optimization algorithms solve the reconfiguration problem by using conventional optimization techniques, for example, OPF by Bender Decomposition [8], mixed-integer convex programming [27, 28], convex models [29], mixed-integer linear programming [30], and mixed-integer second-order cone programming [31].

Nowadays, with the proliferation of photovoltaic systems, many distribution systems could integrate distributed generation (DG), storage systems, and power electronics (STATCOM), so they have to be included in the reconfiguration formulation problem as in Refs. [11, 20, 24, 32–35].

Furthermore, in some papers, multi objective formulation problems are considered. In this sense, formulations include loss minimization and some other function, such as voltage profile enhancement [20, 36–40], load balancing [19, 38, 41, 42], branch current overloads [38], operation cost reduction [43], reliability [32, 44, 45], and outage costs [46].

1.2 Reactive power compensation dispatch

As pointed out before, reactive power compensation (RPC) dispatch is the second way of reducing distribution system electrical losses, so a common objective function is electrical loss minimization either at the planning or operation stage. Furthermore, this objective function is nonlinear and convex, permitting its reduction by sequential commutation of capacitor banks to find one point where its value is minimal until the next capacitor bank commutation causes an electrical loss increment again [47].

It is important to note that the joint application of reconfiguration and RPC strategies allows for obtaining lower losses than either separated so that methods that have been developed involve both strategies [17, 48].

From the planning perspective, loss minimization may be reached by solving an optimal reconfiguration and allocation capacitor problem [33, 34, 49–53], which is combinatorial and nonlinear, so its solution has been proposed by using heuristic methods, metaheuristic methods, and mathematical optimization methods [54].

On the other hand, in an operation environment, loss minimization may be achieved by solving a loss minimization problem by joint reconfiguration and RPC dispatch, which is carried out with capacitor banks already installed with the capacity to be managed from the distribution control center [55]. However, due to the emerging concept of distribution control centers, solutions methods involving reconfiguration and RPC dispatch in real-time are few [54] and are based on ordinal optimization theory [56], parallel metaheuristic [57], multiagent system [55], analytical partitioning method [58], modified binary gray wolf [35], and robust optimization model [59].

This chapter presents two methodologies for solving the reconfiguration and RPC dispatch. Both methodologies are formulated as two-stage reconfiguration and RPC dispatch problems. In the first stage, network reconfiguration is carried out to find a set of feasible radial configurations with the lowest losses, satisfying voltage, and overload restrictions. In the second stage, an RPC dispatch is applied to each feasible network configuration by connecting capacitor blocks successively until all available reactive capacity has been used or a specified loss tolerance has been reached. Finally, analysis of switching each capacitor block is carried out using voltage/shunt reactive compensation linear sensitivities to make a relatively low computational work during the process.

2. Network reconfiguration and reactive power compensation formulation problems

This section describes the methodology proposals for reconfiguration and RPC dispatch. Firstly, considerations for modeling and problem formulation are defined, bearing in mind the electrical loss minimization as the objective function, subject to various restrictions observed during the overall solution process.

To develop the problem formulation, it is necessary to make the next considerations, which are as follows—(i) a three-phase distribution system is operating under phase balance so that the network model is defined by the positive sequence circuit of its components; (ii) the distribution system may be supplied by one or more substations; (iii) RPC dispatch is realized by already installed capacitor banks and they can be commutated remotely from the distribution control center; (iv) capacitor banks only have either a commercial capacity of 300, 600, or 900 kVAr.

2.1 Problem formulation

The problem of electrical loss minimization in distribution systems considering reconfiguration and RPC dispatch can be formulated as follows:

$$Min\ P_{loss} = \sum_{k \in F_B} \frac{r_k}{r_k^2 + x_k^2} \left(V_i^2 + V_m^2 - 2 V_i V_m \cos\left(\theta_i - \theta_m\right)\right) \quad (1)$$

$$S.to\ P_{Gi} - P_{Di} - V_i \sum_{m \in i} V_m Y_{im} \cos(\theta_i - \theta_m - \gamma_{im}) = 0, \quad \forall i \in N \tag{2}$$

$$Q_{Gi} - Q_{Di} - V_i \sum_{m \in i} V_m Y_{im} \sin(\theta_i - \theta_m - \gamma_{im}) = 0, \quad \forall i \in N \tag{3}$$

$$V_i^{Min} \le V_i \le V_i^{Max}, \quad \forall i \in N \tag{4}$$

$$S_k \le S_k^{Max}, \quad \forall k \in F_B \tag{5}$$

$$S_{out,t} \le S_{out,t}^{Max}, \quad \forall t \in T_S \tag{6}$$

$$0 \le b_c \le b_c^{Max}, \quad \forall c \in N_C \tag{7}$$

$$|F_B| = |N| - 1 \tag{8}$$

$$\left\{ (i_1, m_1) \cup (i_2, m_2) \cdots \cup \left(i_{|F_B|}, m_{|F_B|} \right) \right\} = N \tag{9}$$

where b_c represents a capacitor bank susceptance, F_B is the feeder section set, r_k and x_k are the series resistance and reactance of the kth feeder section, respectively. S_k is the apparent power flowing on the kth feeder section and S_k^{Max} is the rating of the kth feeder section, T_S is the set of supply transformers in the distribution system. V_i and θ_i are the voltage magnitude and angle of complex voltage $\overline{V}_i$ at bus i, and V_i^{Max}, V_i^{Min} are the maximum and minimum voltage magnitudes at bus i, respectively. Y_{im} and γ_{im} are the magnitude and angle of the complex nodal admittance associating busses i and m.

The objective function (1) accounts for the distribution system losses. The decision variables are the voltage magnitudes and angles of complex nodal voltages because their values define the losses at each feeder section. These decision variables are to be modified using reconfiguration and RPC dispatch.

Constraints (2) and (3) are the active (P) and reactive power (Q) balances at each bus $i \in N$. Constraint (4) considers that all nodal voltage magnitudes should remain within limits. Constraint (5) takes care of apparent power flow that does not reach a value above its maximum limit through each feeder section. Because load transfers are possible when the distribution system is being reconfigured, constraint (6) is necessary for imposing a maximum limit, $S_{out,t}^{Max}$, to the apparent power flow, $S_{out,t}$, from the supply substation transformer t to distribution feeders. Constraint (7) refers to every capacitor bank c of the set N_C, denoted in terms of its susceptance; it may have a zero value when disconnected and connected by one or more steps (blocks) until it reaches its maximum value $b_c^{\max}$. Finally, constraints (8) and (9) refer to guarantee radiality and maximum spanning tree of the network, which are based on the concept of set cardinality, Constraint (8) indicates that cardinality of F_B, $|F_B|$, should be equal to the cardinality of N, $|N|$, minus 1, while Constraint (9) guarantees that the radial network is a spanning tree, due that the left-hand set should be equal to N.

This general formulation is cast as a nonlinear and combinatorial problem. It can be solved by heuristic, metaheuristic, mathematic optimization, and combinations of those mentioned above. In general, a power flow should be realized at each solution step. Then, it is important that solution methods can find the optimal configuration rapidly to reduce the number of power flow simulations and computing time. Also, computational efficiency can be improved if the power flow algorithm is efficient, like those developed for solving radial distribution power flows [54]. However, these algorithms are not useful for the proposal presented in this chapter due to sensitivity calculations.

2.2 Power flow and sensitivity calculations

Because sensitivity calculations are straightforward from the linearized power flow model solved in each iteration of the Newton–Raphson (NR) method, this is the algorithm used for the methodologies developed for solving the reconfiguration and RPC dispatch to minimize electrical distribution losses.

Equation sets (2) and (3) represent the power flow problem, which is solved iteratively applying the NR method by formulating and solving the next linear equation set expressed in the compact form [60]:

$$\begin{bmatrix} \Delta P \\ \\ \Delta Q \end{bmatrix}^{(l)} = \begin{bmatrix} \frac{\partial P}{\partial \theta} & \frac{\partial P}{\partial V} V \\ \frac{\partial Q}{\partial \theta} & \frac{\partial Q}{\partial V} V \end{bmatrix}^{(l)} \begin{bmatrix} \Delta \theta \\ \\ \frac{\Delta V}{V} \end{bmatrix}^{(l)} \tag{10}$$

Once solved (10), nodal voltage angles and magnitudes are updated as follows:

$$\theta_i^{(l+1)} = \theta_i^{(l)} + \Delta \theta_i^{(l)}, \quad \forall i \in N \tag{11}$$

$$V_i^{(l+1)} = V_i^{(l)} + [\Delta V_i / V_i]^{(l)} V_i^{(l)} \quad \forall i \in N \tag{12}$$

where l is the iteration number, this process continues until a convergence tolerance is accomplished. In Eq. (10), the partial-derivative matrix is the Jacobian matrix and should be calculated at iteration l. Let $\left[\theta^{(base)} \quad V^{(base)}\right]^T$, the power flow problem solution once that iterative process has been finished, where T indicates transposed. This vector solution represents the electrical system base case state, that is, the operating point defined by load, supply, electric network power flows, and losses. For sensitivity calculations, it is known as the base case. Suppose that this equilibrium point is perturbed with the commutation of a capacitor bank c installed at bus j, denoted by Δb_{cj}. Therefore, changes in the state vector can be calculated by constructing the next linear equation set [61]:

$$\begin{bmatrix} \frac{\partial P}{\partial \theta} & \frac{\partial P}{\partial V} \\ \\ \frac{\partial Q}{\partial \theta} & \frac{\partial Q}{\partial V} \end{bmatrix} \begin{bmatrix} \frac{\Delta \theta}{\Delta b_c} \\ \\ \frac{\Delta V}{\Delta b_c} \end{bmatrix} = - \begin{bmatrix} 0 \\ \vdots \\ 0 \\ -V_j^2 \\ 0 \\ \vdots \\ 0 \end{bmatrix} \tag{13}$$

The only nonzero entry in the right-hand vector indicates that capacitor bank c is connected at bus j. Vector $\left[\Delta\theta/\Delta b_{cj} \quad \Delta V/\Delta b_{cj}\right]^T$ is known as the relative sensitivity vector between the state base case vector $\left[\theta^{(base)} \quad V^{(base)}\right]^T$ and the perturbation scalar value Δb_{cj}. The Jacobian matrix in Eq. (13) may be taken from Eq. (10) formed, ordered, and factorized in the last NR iteration, but noting that submatrices $[(\partial P/\partial V)V]$ and $[(\partial Q/\partial V)V]$ in Eq. (10) are divided by their corresponding V. After the sensitivity calculations, a new state vector can be obtained, by using the next equations, once that Δb_{cj} has been defined in terms of kVAr:

$$\theta_i^{(new)} = \theta_i^{(base)} + \left(\Delta\theta_i/\Delta b_{cj}\right)\Delta b_{cj}, \quad \forall i \in N \tag{14}$$

$$V_i^{(new)} = V_i^{(base)} + \left(\Delta V_i/\Delta b_{cj}\right)\Delta b_{cj}, \quad \forall i \in N \tag{15}$$

With these new values, power flows and distribution losses are recalculated to know if the capacitor block Δb_{cj} connection causes a decrease or increase in the objective function, that is, total distribution system losses. If the distribution system has installed a set of capacitor banks, denoted as $\{b_{c1}, \dots, b_{cN_c}\}$, Eq. (13) should be solved each time one capacitor bank or block is connected, which seems to be a high computational work; however, the Jacobian matrix in this equation remains constant and factored, so that, each solution of Eq. (13) requires only a forward and a backward substitution process, which represent much lower computational effort than the one related with a complete power flow calculation.

A flow chart of the power flow algorithm based on the NR method is shown in **Figure 1**, while sensitivity calculation is shown in **Figure 2**.

Notes about the **Figure 1**:

a. $\Delta P_{\max}$ and $\Delta Q_{\max}$ are defined by selecting the greatest values of $\Delta P_i^{(l)}$ and $\Delta Q_i^{(l)}$, $\forall i \in N$.

b. Expression $i \neq s$ signifies a node designated as slack, which contributes with all active and reactive powers plus electrical distribution losses for always keeping the balance power of restrictions (2) and (3). This node represents the supply point for all the system feeders in distribution systems. Furthermore, the slack node maintains constant values of V_s and θ_s during the power flow solution process. V_{spec} is a value around 1.0 per unit value.

c. $lmax$ is the maximum number of iterations for the NR method, and currently, NR converges to the solution in a few iterations because of its quadratic convergence characteristic.

d. Convergence tolerances tol_P and tol_Q are defined as 0.001 or 0.0001.

e. Post-iterative calculations refer to the determination of power flows through each feeder section and electrical losses in the distribution system as follows:

$$P_{im} = V_i\left[V_i g_{im} - V_m y_{im}\cos\left(\theta_i - \theta_m - \varphi_{im}\right)\right] \tag{16}$$

$$P_{mi} = V_m\left[V_m g_{im} - V_i y_{im}\cos\left(\theta_m - \theta_i - \varphi_{im}\right)\right] \tag{17}$$

where complex series admittance of feeder section connecting nodes i and m is calculated in rectangular and polar coordinates by Eqs. (18) and (19), respectively.

$$\bar{y}_{im} = g_{im} + jb_{im} = \frac{r_{im}}{r_{im}^2 + x_{im}^2} + j\frac{x_{im}}{r_{im}^2 + x_{im}^2} \tag{18}$$

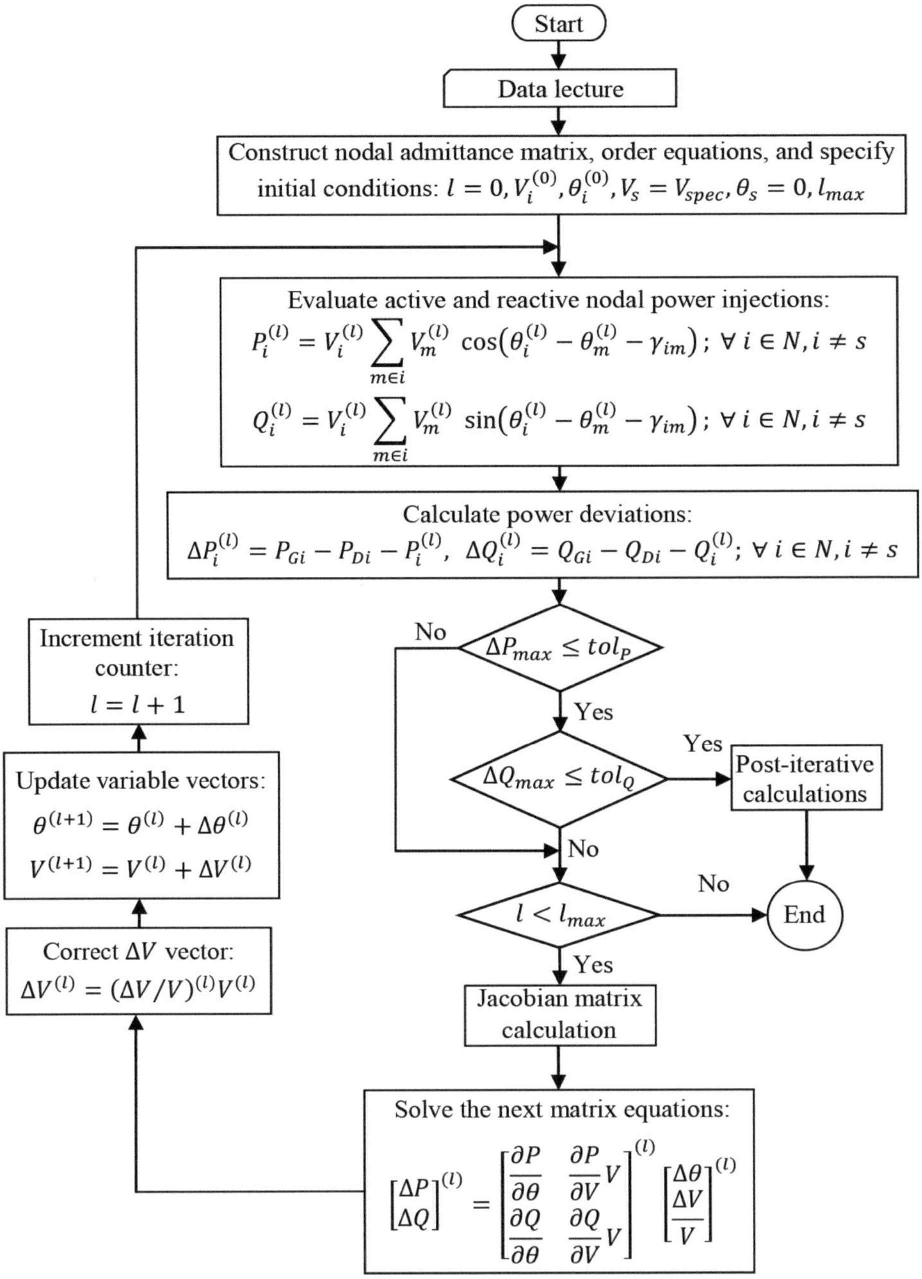

Figure 1.
Flow chart for solving power flow problems by Newton Raphson method.

$$\bar{y}_{im} = y_{im} \angle \varphi_{im}; \ y_{im} \sqrt{g_{im}^2 + b_{im}^2}; \ \varphi_{im} = \tan^{-1}(b_{im}/g_{im}) \tag{19}$$

Active power losses through the feeder section connecting nodes i and m are calculated by Eqs. (16) and (17), which, after some algebraic operations, are expressed as:

$$P_{loss,i-m} = g_{im}\left[V_i^2 + V_m^2 - 2V_i V_m \cos(\theta_i - \theta_m)\right] \tag{20}$$

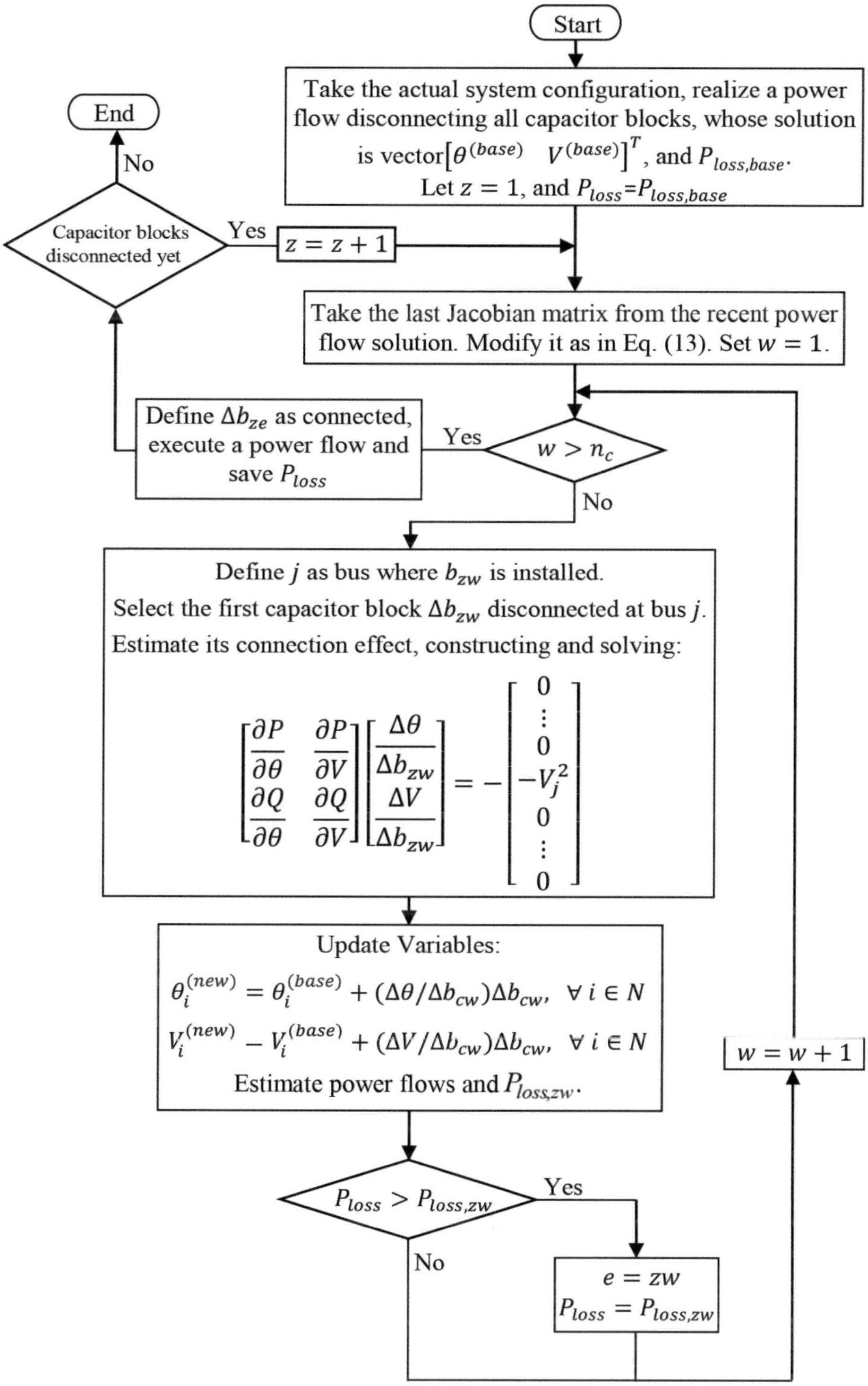

Figure 2.
Flow chart for sensitivity evaluations of capacitor bank connections.

Eq. (20) is the same as Eq. (1) but g_{im} written in terms of r_{im} and x_{im}.

f. In addition, among other post-iterative calculations, sensitivity evaluations when connected to capacitor banks may be included in the next explanation.

2.2.1 Sensitivity Evaluation to Find the Maximum Loss Reduction

By using the power flow solution obtained by the NR algorithm, sensitivities for capacitor bank connections may be calculated straightforwardly. RPC dispatch assumes that the already installed capacitor banks have initially disconnected one, two, three, or more blocks. A power flow is performed to determine the initial distribution losses. After, capacitor blocks will be connected successively. The process calculates sensitivities for each block connection evaluating the new state distribution system, power flows, and distribution losses. Once the first block pertaining to every one of the capacitors installed were connected, their corresponding distribution losses were ranked from the lowest to the highest. The capacitor block associated with the first position is added to a connected set of capacitor blocks. Then, a new power flow is carried out to refresh the Jacobian matrix to continue the sensitivities calculation to evaluate the connection of the next capacitor block remaining as disconnected in each node where capacitors exist in the distribution system. This procedure is carried out until no more capacitor blocks are disconnected or optimal distribution losses are found. The flow chart of **Figure 2** resumes this procedure, where $n_c = |N_C|$, that is, the number of capacitor banks.

To observe the behavior of distribution losses with capacitor bank connections, **Figure 3** shows a distribution feeder whose section parameters are all equal on a per unit basis: $r = 0.1$ and $x = 0.07$ over a base of 10 MVA and 13.8 kV. Node 0 is the supply point with no load, while all the other nodes have a uniform load of 120 kW and 60 kVAr.

The analysis was realized by simulating the connection of a capacitor bank in nodes 1 through 12 to find the optimum capacitor bank and its location. For the sake of clarity, **Figure 4** shows the results only for nodes 7, 8, and 9, which showed lower losses. Note that distribution losses were plotted from 300 kVAr to 750 kVAr, with a linear distribution loss reduction up to 480 kVAr capacitor bank connection in the three nodes, where node 9 presents the lowest distribution losses. After this point, distribution losses become more nonlinear until reaching a minimum value of around 116.5 kW with 620 kVAr in node 9, 116.0 kW with 660 kVAr in node 8, and 116.4 kW with 680 kVAr in node 7. From this analysis, important concluding remarks are as follows—(i) the node with the initial lowest distribution losses could not be the same when the minimum is found after capacitor connections are performed; (ii) minimum loss values reached with capacitor connections are very close to each other, and the same occurs with their kVAr capacities, which are very similar; (iii) the block capacitor connection effect is steadily decreasing over distribution losses until it arrives at a

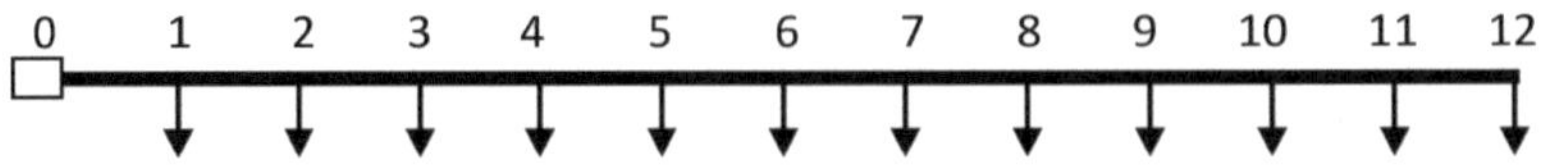

Figure 3.
Distribution system with equal section parameters and uniform load.

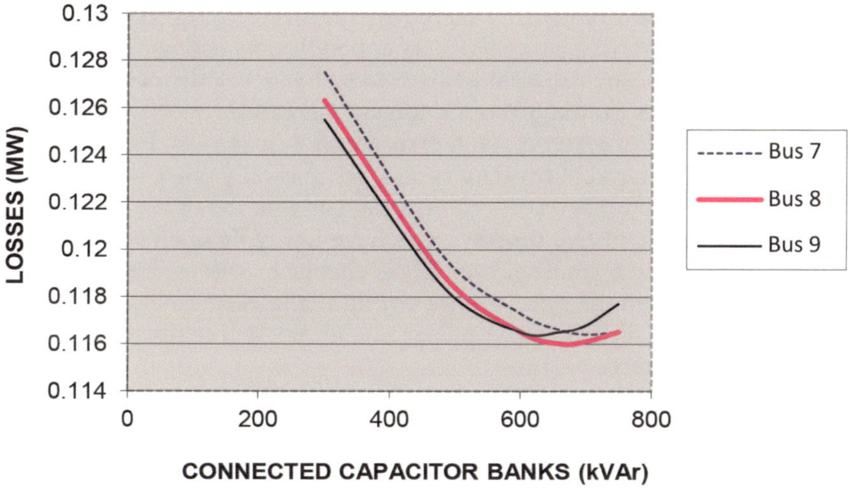

Figure 4.
Distribution electrical losses versus capacitor bank connection.

practically zero value, and after this point, the effect tends to be negative; (iv) lately, it can be considered that the optimal distribution losses can be found with an error below of 1 kW, so that, this tolerance is used as stopping criterium for both methodologies explained next.

3. Methodologies

Formulations proposed in Refs. [5, 47] solve the network reconfiguration problem in distribution systems following a similar strategy. First, the initial radial distribution system is converted to a meshed network by closing all the system switches; then one switch is opened according to the lowest apparent power flow criterion. Once this action is finished, the next switch with the lowest apparent flow is opened, and so on. This process continues until the resulting configuration is completely radial, and its graph is a spanning tree. This algorithm can find the optimal or suboptimal reconfiguration, that is., with the lowest or almost lowest distribution of electrical losses. An alternative strategy consists in looking for the two or more reconfigurations whose switches cause the lower distribution losses instead of searching for only one reconfiguration at each step. This will lead to a feasible reconfiguration set. Furthermore, as explained before, the resultant distribution losses may be further reduced with the application of RPC dispatch to those reconfigurations obtained before, which have minor losses and are very close to each other, as in the example described above. Thus, once the reconfiguration process has finished (Stage 1), the result is a feasible reconfiguration set containing a reduced reconfiguration alternative number, which passes through the RPC dispatch procedure for obtaining the optimal loss reconfiguration (Stage 2).

3.1 Feasible reconfiguration search algorithm

The basic algorithm developed in Ref. [5] looks for only one reconfiguration, which, almost in all cases, yields the one with the lowest distribution losses. In this chapter, this algorithm is modified for searching various reconfigurations based on the selection of the two switches having the least apparent power flow through them.

With a view to clarifying explanation, consider that every distribution system, for simplicity, is operated as a radial network and supplies all the electricity consumers.

In general, distribution systems can be represented by graphs. For example, the distribution system has a radial configuration if there is only one path between the supply and load points. Furthermore, for making possible the optimal operation with the least electrical losses, distribution systems have normally open switches, which can be closed when some operational condition change causes electrical losses to increase significantly, and the same number of normally closed switches should be opened for maintaining distribution system radiality providing that there is not any node isolated from the rest.

Let us assume that switches can be opened or closed from the distribution control center, so the distribution electrical network topology may be updated depending on the prevailing operational conditions in the distribution system.

In terms of graph theory, a graph is formed by branches and links, which connect vertices (nodes), so that if the latter are separated from the graph, the result is a graph with only one trajectory between any two nodes in the graph and the same occurs with a radial distribution system.

An assumption can be made—normally closed switches can be considered branches, while normally open switches are regarded as links. Defining the branch number as B, link number as L, and the extreme points of every switch as nodes, whose number is M, therefore, the tree graph has $M - 1 = B$ branches [62]. Also, it is called a connected tree graph or spanning tree if this graph includes all radial distribution system nodes. Networks whose graphs are not tree graphs but contain all their nodes, that is, they do not have isolated nodes. Therefore, they are just connected graphs, as in the case where one or more of the normally open switches are closed, creating a meshed distribution system.

With the graph information and using a breadth-first search algorithm (BFA), the search process for finding the feasible reconfiguration set (FRS) is described as follows.

The BFA organizes its search in levels by opening some of the normally closed switches from the base case. **Figure 5** illustrates this procedure.

The procedure begins at level 0 with no open switches, denoted as the set $OS = \{\}$. Therefore, considering the two switches, S_1 and S_2, with the lowest apparent power flow, the first level is formed with the sets: $OS = \{S_1, S_2\}$, $FR_1^{(1)} = \{S_1\}$, $FR_2^{(1)} = \{S_2\}$, which are used for the level 2, giving the next sets: $OS = \{S_1, S_2, S_3, S_4\}$, $FR_1^{(2)} = \{S_1, S_3\}$, $FR_2^{(2)} = \{S_1, S_4\}$, $FR_3^{(2)} = \{S_2, S_5\}$, $FR_4^{(2)} = \{S_2, S_6\}$, and so on. The number of levels is L, so that, at the last level, $OS = \{S_1, \ldots, S_R\}$, where R is the number of open switches at this level.

On the other hand, the construction of feasible reconfigurations, which have only open switches, is carried out with the progress informing $FR_1^{(1)}$, $FR_2^{(1)}$, $FR_1^{(2)}$, $FR_2^{(2)}$, $FR_3^{(2)}$, ..., etc. The number of reconfigurations created by brute force is 2^L, but any violation in either constraint (4)–(6), (8), (9), or by the existence of similar reconfigurations during the searching process, may reduce this number substantially at the end of stage 1.

3.2 Methodology 1

The proposal of Methodology 1 is built upon the method described in Refs. [47, 63] to obtain the compensation scheme. This proposal aims to obtain an accurate solution

DOI: http://dx.doi.org/10.5772/intechopen.102820

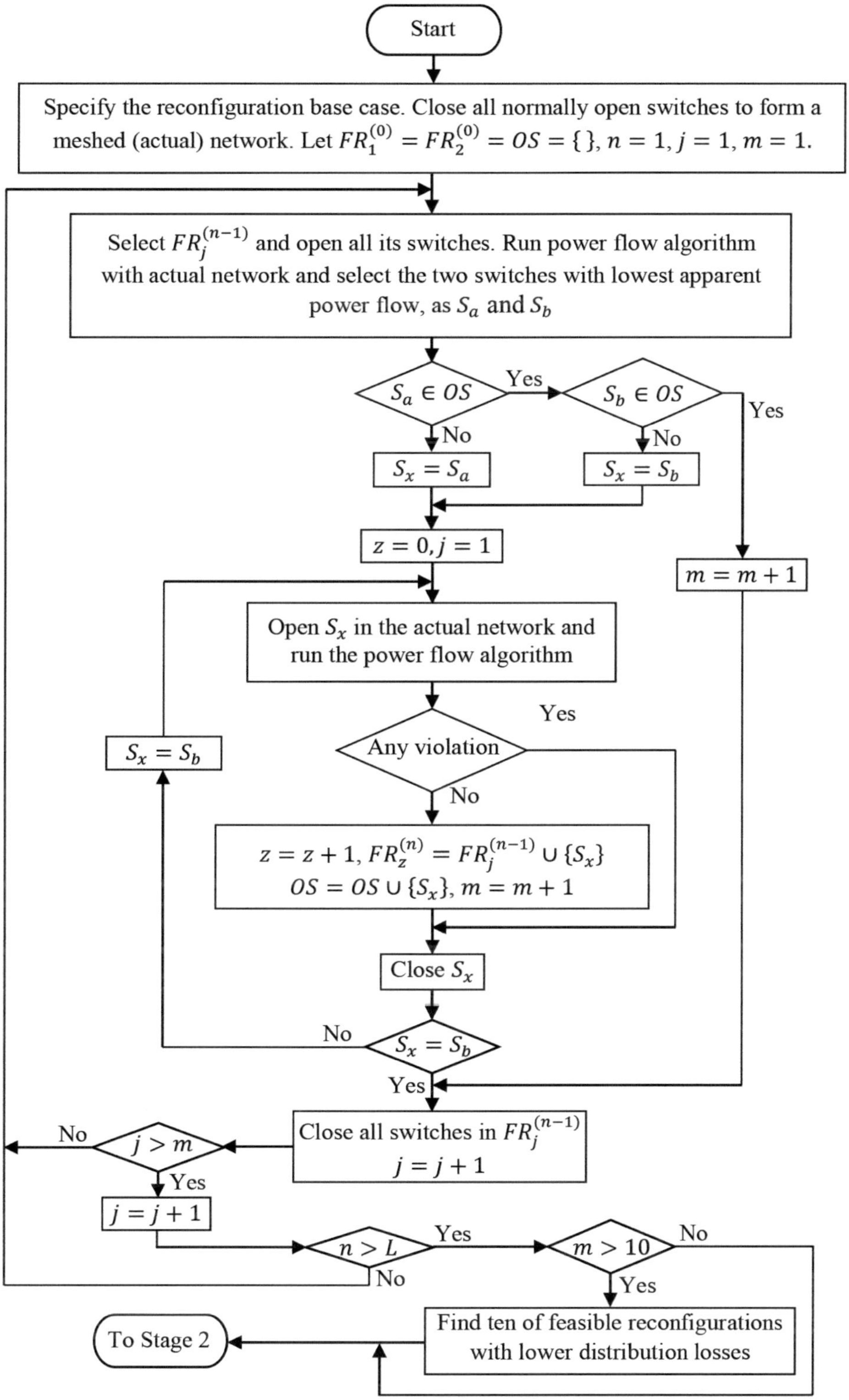

Figure 5.
Flow chart for finding the feasible reconfiguration set.

similar to other methods but tries to maintain a reasonable computational efficiency to be used as a tool for the operation of electrical distribution systems. This proposal looks for the optimal reconfiguration by performing two stages, where the first is related to the search of the FRS using the process illustrated in **Figure 5**. In contrast, the second one realizes the RPC dispatch considering the FRS. The steps of stages 1 and 2 in Methodology 1 are described below.

Stage 1. Determination of a feasible configuration set.

1. All the system switches are closed, forming a meshed system. Also, the *FR* and *OS* sets are initialized as empty sets.

2. A power flow problem is solved, and the two switches with the lowest apparent power flow through them are chosen to form the first level branches, named S_1 and S_2.

3. The selected switches S_1 and S_2 (which do not generate a disconnected graph) are opened separately, and a power flow is executed for each of these resulting network reconfigurations. Suppose the resultant reconfiguration violates some of the constraints (4), (5), and (6). In that case, it is discarded for continuing with the process, and the corresponding switch is closed again for subsequent calculations. Otherwise, go directly to the next step.

4. Open the actual switch at the first level, run the power flow algorithm, and select the next two switches with the lower apparent power flow. Assume that these new switches in the second level are defined as S_3 and S_4, which now will be investigated, and the results will create two branches of the decision-making tree at the third level and so on. This procedure continues until the configuration accomplishes constraints (8) and (9), that is, it is radial and a spanning tree, which will be saved in the feasible reconfiguration set (FRS). Then go to the next step.

5. If the power flow for each of the two switches selected in step 2 was already executed, go to the next step; otherwise, go to step 3.

6. Once the feasible reconfiguration process is terminated, all the feasible reconfigurations are ranked from the lowest to the highest distribution losses. Therefore, after this process, only the first reconfigurations (10 or less, depending on the distribution system size) are considered for performing Stage 2 and Stage 1 finishes.

Stage 2. Reactive power compensation dispatch.

As pointed out before, a node where there is capacitor block(s) installed is consider a feasible system node; also, only capacitor blocks of 300, 600, or 900 kVAr are considered for the RPC dispatch, which is carried out following the next steps:

1. A reconfiguration is selected from the feasible reconfiguration set obtained at the first stage.

2. Independently of the actual state, all the already-installed capacitors in the distribution system are disconnected, and a power flow is executed to compute system losses.

3. A capacitor block is connected to its feasible system node and analyzed by performing a power flow.

4. If there are more feasible nodes to investigate, step 3 is repeated. Otherwise, proceed to the next step.

5. Distribution losses obtained by connecting a capacitor block in each feasible node are compared. The capacitor block that obtains a greater loss reduction is permanently connected.

6. A power flow is executed with the capacitor block connected. If distribution losses were reduced above 1 kW, go to step 3; otherwise, the process is finished.

3.3 Methodology 2

Methodology 1 may be improved, without affecting its accuracy, by considering for the first and subsequent intermediate levels of the decision-making tree, a limit of no more than 10 possible reconfigurations for reducing computational work in the last levels, because, at that levels, they have less influence over distribution loss reductions when selecting different combinations. In addition, the configurations resulting from this process must have lower losses and be close to each other to be taken into account as feasible reconfigurations at Stage 2.

With the above considerations, by making some changes to Methodology 1, Methodology 2 resulted, whose description is as follows.

The experience gained working with Methodology 1 is that, at the first level, considering only two switches with the lower apparent power flow may lead to suboptimal results when the reconfiguration process finishes. Therefore, in level 1, Methodology 2 includes five switches with the lowest apparent power flow. From this point, Methodology 2 continues normally since computational efficiency degrades if this criterion prevails in the next levels.

On the other hand, to improve the computational efficiency of the first stage, the search space at each level is limited so that the decision-making tree does not grow excessively (which happens if there are many link switches).

Furthermore, an additional computational efficiency improvement is disregarding, from the second level and the next ones, those reconfigurations that do not accomplish a tolerance margin of 3%, based on the difference in losses between the reconfiguration with the lowest losses and all other reconfigurations obtained in the correspondent level.

Finally, in Stage 2, of Methodology 2, investigating the connection of capacitor blocks by sensitivity calculations instead of using the complete power flow algorithm may improve the computational efficiency without losing accuracy.

Performing the previous modifications to Methodology 1, the steps of each stage of methodology 2 are defined as follows.

Stage 1. Determination of a set of feasible reconfigurations.

1. All the system switches are closed, forming a meshed system. Also, the *FR* and *OS* sets are initialized as empty sets.

2. A power flow problem is solved, and the five switches (which do not generate islands when opened) with the lowest apparent power flow through them are chosen.

3. The selected switches are opened separately, and a power flow is executed for each of these resulting network reconfigurations. If the resultant reconfiguration violates any of the constraints (4), (5), or (6), it is discarded, and the corresponding feeder section is closed again to go forward with the process of finding the feasible reconfiguration set. Also, reconfigurations that result in one previously defined as feasible are removed, and their corresponding switch is closed again. If these constraints are not violated, go to the next step.

4. The losses obtained for each different reconfiguration are compared. At the first level of the decision-making tree, the three reconfiguration options with the greatest losses are eliminated so that only two remain.

5. If the actual reconfiguration accomplishes constraints (8) and (9), that is, is radial and a spanning tree, save it in the FRS. Also, if all the reconfiguration alternatives are exhausted, go to step 6; otherwise, go to step 2.

6. All the feasible reconfigurations are ranked considering their distribution losses from the lowest to the highest. The reconfigurations that present a distribution loss difference greater than a tolerance margin of 3% concerning the reconfiguration with the lowest losses are eliminated. The remaining reconfigurations form the final FRS, and they will be passed through the RPC dispatch at Stage 2.

Stage 2. Reactive power compensation dispatch.

Under the same considerations of Stage 2 of Methodology 1, the RPC dispatch is carried out following the next steps:

1. A reconfiguration is selected from the feasible reconfiguration set obtained at the first stage.

2. Independently of the actual state, all the already-installed capacitors in the distribution system are disconnected, and a power flow is executed to compute system losses.

3. A capacitor block is connected to its feasible system node and analyzed by the sensitivity algorithm to estimate the new losses.

4. If there are more feasible nodes to investigate, step 3 is repeated; else, proceed to the next step.

5. Distribution losses obtained by connecting a capacitor block in each feasible node are compared. The capacitor block that obtains a greater loss reduction is declared as permanently connected.

6. A power flow is executed with the capacitor block connected. If distribution losses were reduced above 1 kW, the Jacobian matrix is updated for sensitivity calculations and goes to step 3; otherwise, the process is declared finished.

4. Numerical examples

This section reports numerical examples from two case studies. The proposed methodologies are first applied to the IEEE 16-bus system for illustrative purposes. This benchmark is useful to analyze comprehensively the effects of the novel modeling aspects featured by the proposed methods. Subsequently, the performance of both methodologies on the Taiwan Power Company distribution system is investigated.

4.1 IEEE 16-bus distribution system

In addition to being widely used in the literature, the IEEE 16-node distribution system has a reactive compensation scheme, so it is ideal for observing the behavior of algorithms that perform reconfiguration and RPC dispatch for distribution systems. This system is a three-phase distribution system with three feeders, 16 nodes, 13 load points, 13 normally closed switches, and 3 normally opened switches, and the operation nominal voltage is 12.66 kV. **Figure 6** shows the corresponding single line diagram, where switches S14, S15, and S16 define the initial operative condition of this system. Suppose that the load pattern changes so that nodal voltage and angle profiles are modified, causing distribution losses that may not be optimal with the actual configuration and RPC dispatch.

A set of feasible reconfigurations are obtained at Stage 1. Then, the solution with the lowest losses and the resultant RPC dispatch with capacitor banks is obtained at Stage 2.

First, the three normally open switches are closed; thus, a meshed network is formed. Then, there are three meshes in the system, so the same number of switches must be opened for the system to be radial again. Therefore, the decision-making tree consists of three levels. For level 1, the two switches, S7 and S16, are selected because they have the lowest apparent power flow through them. After, a power flow is

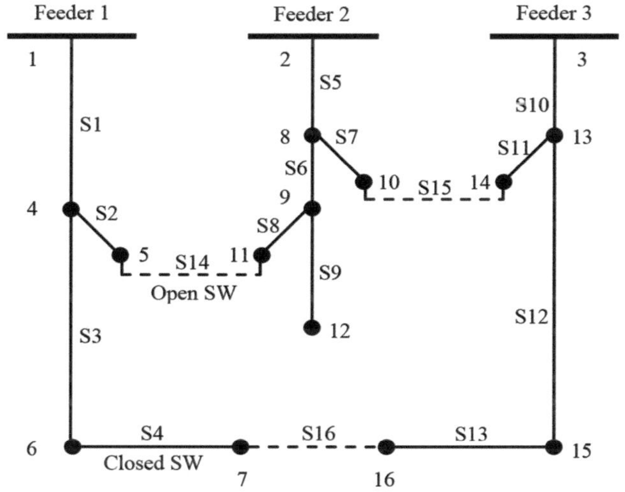

Figure 6.
IEEE 16-node distribution system.

carried out for switches S7, and S16 opened separately, and the next two switches with the lowest apparent power flow are selected.

Figure 7 illustrates the feasible reconfiguration selection process. Note that losses were increasing from the top level to the bottom level. This is due to meshed networks presenting lower losses than radial networks. However, between the five feasible reconfigurations in level 3, $FR_4^{(3)}$ = {S7, S16, S8}, presents the lowest distribution losses, 466.1 kW, which is to be the first candidate to obtain the optimal value at the end of Stage 2 because of the distribution losses given by the other feasible reconfigurations.

On the other hand, three repeated options are eliminated (alternatives with dotted lines), one of them at level 2. Therefore, there is no reason why it must be investigated at lower levels. At the process ending, only five feasible reconfigurations passed to Stage 2.

The resulting reconfiguration options make up a set of feasible configurations, as shown in **Table 1**.

Other methodologies applied to this example selected a reconfiguration with the lowest losses open switches S7, S16, and S8 (FR1), which presents the lowest loss value of 468.33 kW. Therefore, Methodology 1 includes this optimal solution among feasible reconfigurations. Then, stage 2 is applied to know the final losses with RPC dispatch to the five feasible reconfigurations.

For this system, to apply Methodology 2, the five switches with the lowest apparent power flow are selected to be the decision-making tree first level, S7, S16, S4, S8, and S1. The distribution losses obtained with switches S7 and S16 open are less than those obtained when opening the other ones. Therefore, these two switches are selected to begin the decision-making tree and continue the reconfiguration process. The set of feasible reconfigurations is shown in **Table 2**.

Unlike the feasible reconfigurations obtained by Methodology 1, only two configurations resulted from applying Methodology 2. This is because, when considering a tolerance margin of 3%, as indicated in the last step of Stage 1, configurations whose difference in distribution losses concerning the lowest losses obtained is more than 3%, that is, 14.049 kW, are eliminated from the feasible reconfiguration set. The only

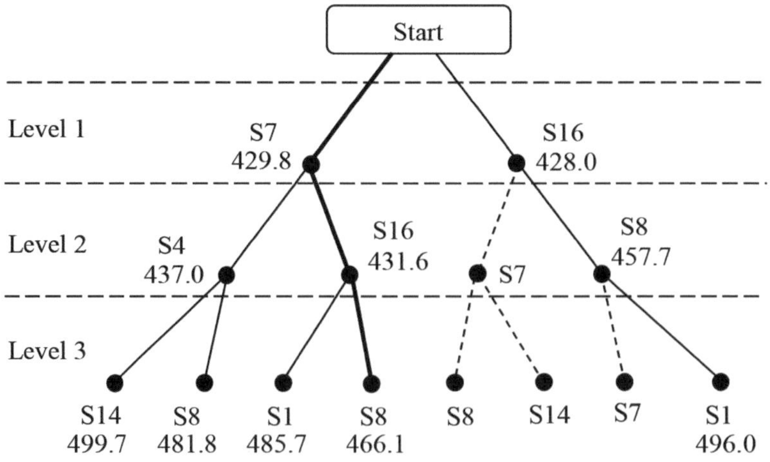

Figure 7.
*Generation of feasible reconfigurations by levels for the system of **Figure 4**.*

Feasible reconfiguration	Open switches	Losses (kW)
FR1	S7, S16, S8	468.33
FR2	S7, S16, S1	485.77
FR3	S7, S4, S8	481.82
FR4	S7, S4, S14	499.78
FR5	S16, S8, S1	496.00

Table 1.
Feasible reconfigurations obtained (first stage) using methodology 1.

Feasible Reconfiguration	Open switches	Losses (kW)
FR1	S7, S16, S8	468.33
FR2	S7, S4, S8	481.82

Table 2.
Feasible configurations found using methodology 2.

	Methodology 1					Methodology 2	
Feasible reconfiguration	FR1	FR2	FR3	FR4	FR5	FR1	FR2
Losses without reactive compensation (kW)	606.6	628.0	627.2	650.5	634.3	606.6	627.2
Losses with reactive compensation (kW)	461.5	480.2	471.9	492.3	479.7	461.5	471.9
Loss reduction (kW)	145.1	147.8	155.3	158.2	154.6	145.1	155.3

Table 3.
Results of stage 2 applying both methodologies.

feasible reconfiguration that complies with this tolerance is the one corresponding to the opening of switches S7, S4, and S8, which corresponds to the FR3 when applying Methodology 1.

The total reactive capacity compensation defined for the original system is 11.4 MVAr. Therefore, 11.4 MVAr will also be used as a limit for the RPC dispatch or until the loss reduction is less than 1 kW. Each capacitor block is 300 kVAr. Hence, 38 blocks can be used to cover the total reactive capacity. The results are shown in **Table 3**.

As can be noted, with reconfiguration FR1, the minimum losses are obtained before and after performing the RPC dispatch. In addition, it is observed that the higher the losses before starting Stage 2 tend to be reduced the more when applying for reactive compensation. However, this fails to change the result of the combination of reconfiguration and RPC dispatch with minimal losses.

The initial loss difference without connected capacitors between FR2 and FR5 configurations is 6.3 kW or 1.003% over FR2. Therefore, if these two reconfigurations changed positions and FR5 obtained lower losses than FR2 with the capacitors connected, it means that a difference of approximately 1% between the reconfiguration of lower losses and the other feasible reconfigurations is not enough to guarantee that the positions according to lower losses remain the same at the end of Stage 2. In addition, the difference between the losses of the FR2 and FR3

Node	Methodology 1	Methodology 2
5	0.6	0.9
6	0.3	0.3
7	1.2	1.2
8	0.6	0.6
9	3.3	3.3
10	0.6	0.9
11	0.3	0.3
12	2.4	2.4
14	0.6	
15	0.6	0.3
16	0.9	1.2
Losses (kW)	461.44	461.45

Table 4.
Resulting RPC dispatch (MVAr) scheme: IEEE 16-bus system.

configurations before applying the RPC dispatch is only 0.8 kW, and after applying it, the difference results in 8.3 kW.

The solution to the problem of reconfiguration and RPC dispatch for this case is to use FR1 with the RPC dispatch scheme shown in **Table 4**.

From **Table 4**, it can be observed that the difference between methodologies is minimal both in the RPC dispatch scheme and in the final losses. This means that it is adequate to replace the power flow simulations with sensitivity calculations when evaluating losses with the connection of each block of the capacitor complete set.

Methodology 1 determines a different location of capacitor banks than Methodology 2, but the same amount of reactive compensation (11.4 MVAr) is used.

Figure 8 depicts the voltage magnitudes resulting from the base case and Methodologies 1 and 2. Note that, in general, the voltage profile is elevated with respect base case. This is because values of 1 p.u. for nodes 1, 2, and 3 are supply points (see **Figure 6**), and their voltages always are constant. On the other hand, nodes 8, 10, and 11 present the lowest values around 0.97 p.u. and below, whereas the highest voltage magnitude values are presented in nodes 12, 13, 14, 15, and 16. This is because these nodes initially have capacitors connected with higher capacities than those obtained with the RPC dispatch. This situation illustrates that it is better for distribution loss reduction to have deployed capacitor banks in various nodes instead of concentrating them in a few nodes. Finally, it is worth observing the voltage magnitude scale in the vertical axis, and it can be seen that the voltage magnitude differences shown are relatively small.

4.2 Taiwan power company distribution system

The Taiwan Power Company (TPC) distribution system is a three-phase system, 11.4 kV, 11 feeders, 83 normally closed switches, and 13 normally open tie switches. **Figure 9** depicts the network diagram; dotted lines represent normally open tie switches.

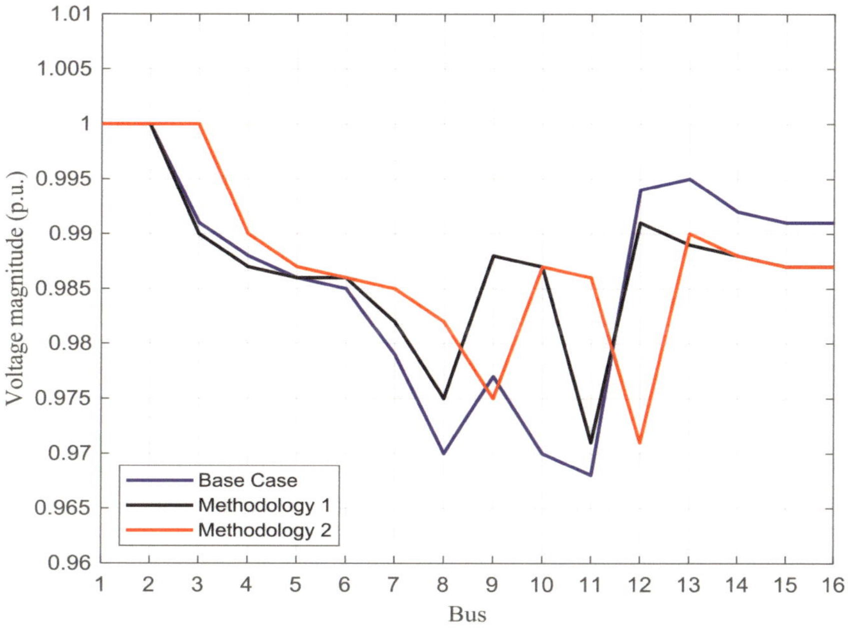

Figure 8.
IEEE 16-bus distribution system voltage magnitudes.

As there are 13 tie-lines, 13 levels for the decision-making tree determine the feasible configurations. With Methodology 1, only the investigation in the last level resulted in 682 feasible reconfigurations. Performing this number of power flow simulations only at the last level's decision-making tree is not convenient for finding FRS. Moreover, for distribution systems with many switches, the efficiency of the proposed methodologies could be similar or even worse than that of almost metaheuristic methods normally used for planning. This is why only 10 configurations with the lower distribution losses are selected, as pointed out before for Stage 1 of both methodologies. Keeping in mind this feature, the result for the current distribution system is shown in **Table 5**, after Stage 1 was realized.

The FRS obtained by Methodology 1 is almost completely different from that resulting from Methodology 2; only three FRs are equal: FR2-FR2, FR5-FR6, and FR9-FR10 defined by methodologies 1 and 2, respectively. Also, the configuration with the minor losses of Methodology 1 reports lower losses than that of Methodology 2. This is because, during the process, the combination of open switches that leads to the configuration of lower losses for Methodology 1 is eliminated when using Methodology 2, since at one level, the limit of 10 possible combinations is exceeded, and that option has higher losses than the 10 with which the process is continued. Despite this, configurations with similar losses were determined since the lowest loss FR obtained by Methodology 1 is 587.2 kW. The one obtained by Methodology 2 is 588.0 kW (the difference is only 0.8 kW or 0.18%). Finally, note that all 10 FRs cause a decrease in electrical losses compared with FR0, denoted as Base.

On the other hand, the combinations formed in the decision-making tree (from the first level to the last) using Methodology 1 are 1135. In the case of Methodology 2, there are only 160 combinations. Because a power flow simulation must be performed for each combination, the difference in computational work between both

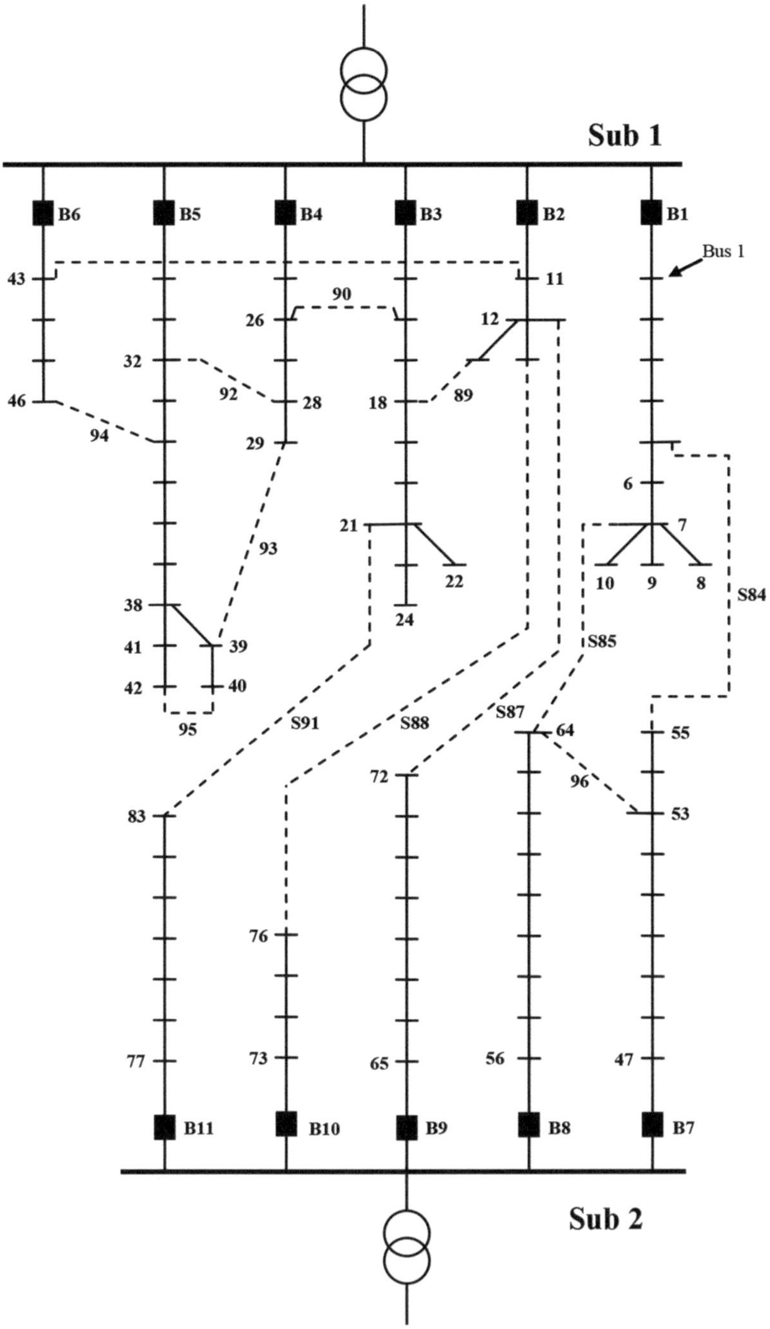

Figure 9.
TPC distribution system.

methodologies applied to the TPC distribution system is too great (975 power flows). However, in general, it may be compensated because there is a small difference between the accuracy of both methods.

The effect of connecting capacitors to each feasible configuration in terms of total distribution losses is reported in **Table 6**. It should be noted that methodologies 1 and 2

FR	Methodology 1		Methodology 2	
	Open switches	**Losses**	**Open switches**	**Losses**
Base	89, 90, 92, 93, 94, 95, 96, 84, 85, 87, 88, 98, 91	343.5 + 254.0 = 599.5	89, 90, 92, 93, 94, 95, 96, 84, 85, 87, 88, 98, 91	343.5 + 254.0 = 599.5
FR1	42, 89, 55, 33, 13, 82, 29, 7, 98, 64, 92, 71, 90	356.2 + 235.3 = 591.5	42, 89, 55, 33, 13, 83, 93, 7, 98, 64, 71, 92	348.9 + 239.3 = 588.2
FR2	42, 89, 55, 33, 13, 83, 29, 7, 98, 64, 92, 71, 90	350.9 + 239.3 = 590.2	42, 89, 55, 33, 13, 83, 29, 7, 98, 64, 92, 71, 90	350.9 + 239.3 = 590.2
FR3	42, 89, 55, 34, 13, 82, 29, 7, 98, 64, 92, 71, 90	348.7 + 239.3 = 588.0	42, 89, 55, 34, 13, 83, 93, 7, 98, 64, 92, 71, 90	347.8 + 239.3 = 587.3
FR4	42, 89, 55, 34, 13, 82, 93, 7, 98, 64, 92, 71, 90	354.1 + 235.3 = 589.4	95, 89, 55, 33, 13, 83, 29, 7, 98, 64, 92, 71, 90	350.8 + 239.3 = 590.1
FR5	42, 89, 55, 34, 13, 83, 29, 7, 98, 64, 92, 71, 90	348.7 + 239.5 = 588.0	42, 89, 55, 34, 13, 83, 29, 7, 98, 64, 32, 71, 90	354.5 + 239.3 = 593.8
FR6	42, 89, 55, 34, 13, 83, 93, 7, 98, 64, 92, 71, 90	347.8 + 239.5 = 587.3	42, 89, 55, 34, 13, 83, 29, 7, 98, 64, 92, 71, 90	348.7 + 239.3 = 588.0
FR7	95, 89, 55, 34, 13, 82, 29, 7, 98, 64, 85, 71, 90	354.4 + 235.3 = 589.7	42, 89, 55, 34, 13, 83, 29, 7, 98, 64, 71, 32, 90	355.2 + 239.3 = 594.5
FR8	95, 89, 55, 34, 13, 82, 93, 7, 98, 64, 92, 71, 90	353.3 + 235.3 = 588.6	42, 89, 55, 33, 13, 83, 93, 7, 98, 64, 92, 71, 90	349.1 + 239.3 = 588.4
FR9	95, 89, 55, 34, 13, 83, 29, 7, 98, 64, 92, 71, 90	349.0 + 239.3 = 588.3	42, 89, 55, 33, 13, 83, 93, 7, 98, 64, 71, 32, 90	353.2 + 239.3 = 592.5
FR10	95, 89, 55,34, 13, 83, 93, 7, 98, 64, 92, 71, 90	347.9 + 239.3 = 587.2	95, 89, 55, 34, 13, 83, 29, 7, 98, 64, 92, 71, 90	349.0 + 239.3 = 588.3

Table 5.
Ten feasible configurations with the lowest distribution losses (kW).

FR	Methodology 1				Methodology 2			
	Before RPC dispatch		**After RPC dispatch**		**Before RPC dispatch**		**After RPC dispatch**	
	Losses	**Rank**	**Losses**	**Rank**	**Losses**	**Rank**	**Losses**	**Rank**
FR1	591.5	10	439.7	10	588.2	3	437.5	3
FR2	590.2	9	438.9	8	590.2	7	438.9	8
FR3	588.0	4	437.9	5	587.1	1	437.3	1
FR4	589.4	7	438.9	9	590.1	6	438.8	7
FR5	588.0	3	437.9	4	593.8	9	439.4	9
FR6	587.3	2	437.3	2	588.0	2	437.9	5
FR7	589.7	8	438.8	7	594.5	10	439.5	10
FR8	588.6	6	438.2	6	588.4	5	437.5	2
FR9	588.3	5	437.8	3	592.5	8	438.5	6
FR10	587.2	1	437.2	1	588.3	4	437.8	4

Table 6.
Distribution losses (kW) and rank with the RPC dispatch for the 10 FRs.

Node	Reactive Power Compensation (MVAr)	
	Methodology 1	Methodology 2
6	1.2	1.2
21	1.8	1.8
28	1.8	1.8
37	0.9	0.9
51	2.1	2.1
63	0.3	0.3
71	2.1	2.1
75	1.5	1.5
81	1.2	1.2
Total compensation (MVAr)	12.9	12.9
Losses (kW)	437.2	437.3

Table 7.
Resulting RPC dispatch (MVAr) scheme: TPC distribution system.

reached practically the minimum electrical loss level with a difference of only 0.1 kW, but, as pointed out before, Methodology 2 is much more efficient than Methodology 1.

The RPC dispatch scheme with Methodology 1 obtained the lowest distribution losses using FR4, whereas Methodology 2 applied RPC dispatch to FR3. Both solutions are reported in **Table 7**. Note that total electrical distribution losses decreased from 587.2 kW without RPC dispatch to 437.2 kW with RPC dispatch applying Methodology 1, that is, a reduction of 150 kW or 25.54%.

Figure 10 depicts the voltage profile for the initial configuration and the proposed methodologies. It can be noticed that the results obtained by the proposed methodologies are quite similar. Both methodologies find a better voltage profile since the minimum voltage is 0.9642, p.u. whereas in the base case, it reaches 0.9466 p.u.

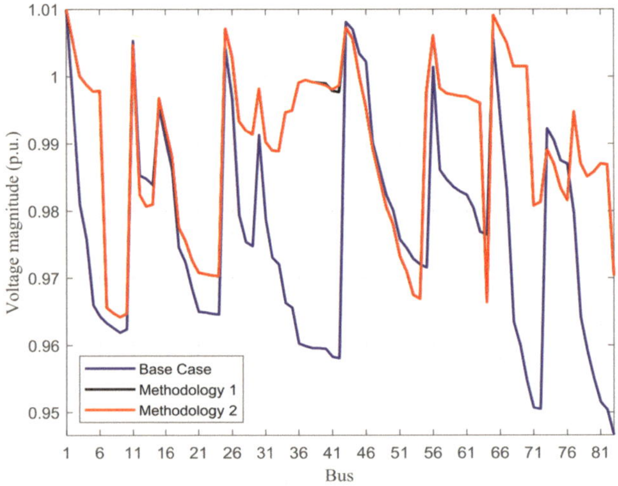

Figure 10.
TPC distribution system voltage magnitudes.

DOI: http://dx.doi.org/10.5772/intechopen.102820

5. Conclusions

Applying reconfiguration and RPC dispatch with capacitor banks simultaneously, even when different formulations are used to solve both problems, allows for a greater reduction in electrical distribution losses. Furthermore, if reconfiguration and RPC dispatch are executed sequentially, the optimal solution may be obtained.

The operation of electrical distribution systems with already installed reactive compensation requires an RPC dispatch to reduce electric losses or improve voltage profiles. Therefore, only nodes with already connected capacitors should be considered for the study. In this regard, the efficiency of the proposed methodologies is increased due to the fact that the search space is reduced and with it the number of necessary operations.

Sensitivity analysis instead of power flow increases the computational efficiency to calculate changes in capacitive compensation.

Both methodologies are a competitive option for distribution loss optimization because of their relatively low computational work. Also, Methodology 2 has higher computational efficiency and accuracy comparable with Methodology 1.

From the results of Stage 1 with the two distribution systems analyzed, it is convenient to limit the FRS to 10 or fewer and consider only those that meet a 3% tolerance with respect to that of lower losses, since only those that report the lowest losses are most likely to be part of the solution at the end of the capacitive compensation process.

When RPC dispatch is performed, a minimum tolerance margin is defined to reduce losses for planning or operational decisions on the distribution system to increase computational efficiency. This loss reduction margin (which also improves the voltage profile of the system in general) must justify the investment made in the capacitor banks and in their connection to the busses that require it.

Acknowledgements

This work was supported by the Tecnologico Nacional de México under Project 13322.21-P.

Author details

Ulises Tovar Ramírez[1], José Horacio Tovar Hernández[2] and
Guillermo Gutiérrez Alcaraz[2*]

1 National Energy Control Center, Mexico City, Mexico

2 National Tehcnological Institute of Mexico/I.T. Morelia, Morelia, Mich, México

*Address all correspondence to: guillermo.ga@morelia.tecnm.mx

References

[1] Civanlar S, Grainger JJ, Yin H, Lee SSH. Distribution feeder reconfiguration for loss reduction. IEEE Transactions on Power Delivery. 1988;**3**:1217-1223. DOI: 10.1109/61.193906

[2] Braz HD, de Souza BA. Distribution network reconfiguration using genetic algorithms with sequential encoding: Subtractive and additive approaches. IEEE Transactions on Power Systems. 2011;**26**:582-593. DOI: 10.1109/TPWRS.2010.2059051

[3] Wu W-C, Tsai M-S. Application of enhanced integer coded particle swarm optimization for distribution system feeder reconfiguration. IEEE Transactions on Power Delivery. 2011;**26**:1591-1599. DOI: 10.1109/TPWRS.2010.2094212

[4] Bouchard DE, Salama MMA, Chikhani AY. Algorithms for distribution feeder reconfiguration. In: 1996 Canadian Conference on Electrical and Computer Engineering. London: IEEE; 1996. pp. 794-797. DOI: 10.1109/CCECE.1996.548272

[5] Shirmohammadi D, Wayne HH. Reconfiguration of electric distribution networks for resistive line losses reduction. IEEE Transactions on Power Delivery. 1989;**4**:1492-1498. DOI: 10.1109/61.25637

[6] Pahwa A. Role of distribution automation in restoration of distribution systems after emergencies. Power Engineering Society Summer Meeting. 2001;**2001**:1-2. DOI: 10.1109/PESS.2001.970136

[7] Khodr HM, Martínez CJ, Matos MA. Distribution systems reconfiguration based on OPF using benders decomposition. IEEE Transactions on Power Delivery. 2009;**24**:2166-2176. DOI: 10.1109/TPWRD.2009.2027510

[8] Mishra S, Das D, Paul S. A comprehensive review on power distribution network reconfiguration. Energy Systems. 2017;**8**:227-284. DOI: 10.1007/s12667-016-0195-7

[9] Zin AAM, Ferdavani AK, Khairuddin AB, Naeini MM. Reconfiguration of radial electrical distribution network through minimum-current circular-updating-mechanism method. IEEE Transactions on Power Systems. 2012;**27**: 968-974. DOI: 10.1109/TPWRS.2011.2174258

[10] Syahputra R, Robandi I, Ashari M. Reconfiguration of distribution network with distributed energy resources integration using PSO algorithm. Telkomnika. 2015;**13**:759-766. DOI: 10.12928/telkomnika.v13i3.1790

[11] Thakur T. Application of tabu-search algorithm for network reconfiguration in radial distribution system. In: 2006 International Conference on Power Electronic, Drives and Energy Systems. London: IEEE; 2006. pp. 1-4. DOI: 10.1109/PEDES.2006.344305

[12] de Andrade BAR, Ferreira NR. Simulated annealing and tabu search applied on network reconfiguration in distribution systems. In: 2018 Simposio Brasileiro de Sistemas Eletricos (SBSE) IEEE: London; 2018. pp. 1-6. DOI: 10.1109/SBSE.2018.8395757.

[13] Cheng HC, Kou CC. Network reconfiguration in distribution systems using simulated annealing. Electric Power Systems Research. 1990;**29**: 239-246. DOI: 10.1016/0378-7796(94)90018-3

[14] Jeon YJ, Kim JC, Kim JO, Shin JR, Lee KY. An efficient simulated annealing algorithm for network reconfiguration in large-scale distribution systems. IEEE Transactions on Power Delivery. 2002; **17**:1070-1078. DOI: 10.1109/TPWRD. 2002.803823

[15] Chiou J-P, Chang C-F, Su C-T. Variable scaling hybrid differential evolution for solving network reconfiguration of distribution systems. IEEE Transactions on Power Systems. 2005;**20**:668-674. DOI: 10.1109/ TPWRS.2005.846096

[16] Chang C-F. Reconfiguration and capacitor placement for loss reduction of distribution systems by ant colony search algorithm. IEEE Transactions on Power Systems. 2008;**23**:1747-1755. DOI: 10.1109/TPWRS.2008.2002169

[17] Abdelaziz AY, Elkhodary SM, Osama RA. Distribution networks reconfiguration for loss reduction using the Hyper Cube Ant Colony Optimization. In: The 2011 International Conference on Computer Engineering & Systems. London: IEEE; 2011. pp. 79-84. DOI: 10.1109/ICCES.2011.6141016

[18] Rama Rao PVV, Sivanagaraju S. Radial distribution network reconfiguration for loss reduction and load balancing using plant growth simulation algorithm. International Journal on Electrical Engineering and Informatics. 2010;**2**:266-277

[19] Rajaram R, Satish Kumar K, Rajasekar N. Power system reconfiguration in a radial distribution network for reducing losses and to improve voltage profile using modified plant growth simulation algorithm with distributed generation (DG). Energy Reports. 2015;**1**:116-122. DOI: 10.1016/j. egyr.2015.03.002

[20] Mam M, Leena G, Saxena NS. Distribution network reconfiguration for power loss minimization using bacterial foraging optimization algorithm. International Journal of Engineering and Manufacturing. 2016;**2**:18-32. DOI: 10.5815/ijem.2016.02.03

[21] Reddy AS, Reddy MD, Reddy MSK. Network reconfiguration of distribution system for loss reduction using GWO algorithm. International Journal of Electrical and Computer Engineering (IJECE). 2017;**7**(6):3226-3234. DOI: 10.11591/ijece.v7i6.pp3226-3234

[22] Yodphet D, Onlam A, Siritaratiwat A, Khunkitti P. Electrical distribution system reconfiguration for power loss reduction by the Salp Swarm algorithm. International Journal of Smart Grid and Clean Energy. 2019;**8**:156-163. DOI: 10.12720/sgce.8.2.156-163

[23] Duong TL, Nguyen TT. Network reconfiguration for an electric distribution system with distributed generators based on symbiotic organism search. Engineering, Technology & Applied Science Research. 2019;**9**:4925-4932. DOI: 10.48084/etasr.3166

[24] Zubairu FA, Jibril Y, Abubakar AS. Optimal reconfiguration of power distribution network for loss reduction using hybrid cuckoo search algorithm and particle swarm optimization. Journal of Science Technology and Education. 2021;**9**:1-10

[25] Rao RS, Narasimham SVL, Raju MR, Rao AS. Optimal network reconfiguration of large-scale distribution system using harmony search algorithm. IEEE Transactions on Power Systems. 2011;**26**:1080-1088. DOI: 10.1109/TPWRS.2010.2076839

[26] Esmaeilian HR, Fadaeinedjad R, Attari SM. Distribution network

reconfiguration to reduce losses and enhance reliability using binary gravitational search algorithm. In: 22nd International Conference on Electricity Distribution (CIRED 2013). London: IEEE; 2013. pp. 1-4. DOI: 10.1049/cp.2013.0703

[27] Jabr RA, Singh R, Pal BC. Minimum loss network reconfiguration using mixed-integer convex programming. IEEE Transactions on Power Systems. 2012;**27**:1106-1115. DOI: 10.1109/TPWRS.2011.2180406

[28] Taylor JA, Hover FS. Convex models of distribution system reconfiguration. IEEE Transactions on Power Systems. 2012;**27**:1407-1413. DOI: 10.1109/TPWRS.2012.2184307

[29] Novoselnik B, Bolfec M, Bošković M, Baotić M. Electrical power distribution system reconfiguration: Case study of a real-life grid in Croatia. International Federation of Automatic Control (IFAC). 2017;**50**:61-66. DOI: 10.1016/j.ifacol.2017.08.011

[30] Mahdavi M, Romero R. Reconfiguration of radial distribution systems: An efficient mathematical model. IEEE Latin America Transactions. 2021;**19**:1172-1181. DOI: 0000-0002-0454-5484

[31] Amanulla B, Chakrabarti S, Singh SN. Reconfiguration of power distribution systems considering reliability and power loss. IEEE Transactions on Power Delivery. 2012; **27**:918-926. DOI: 10.1109/TPWRD.2011.2179950

[32] Salau AO, Gebru YW, Bitew D. Optimal network reconfiguration for power loss and voltage profile enhancement in distribution systems. Heliyon. 2020;**6**:1-8. DOI: 10.1016/j.eliyon.2020.e04233

[33] Biswas PP, Suganthan PN, Amaratunga GAJ. Distribution network reconfiguration together with distributed generator and shunt capacitor allocation for loss minimization. IEEE Congress on Evolutionary Computation (CEC). 2018; **2018**:1-7. DOI: 10.1109/CEC.2018.8477894

[34] Pamshetti VB, Singh S, Singh SP. Combined impact of network reconfiguration and volt-VAR control devices on energy savings in the presence of distributed generation. IEEE Systems Journal. 2020;**14**:995-1006. DOI: 10.1109/JSYST.2019.2928139

[35] Li P, Zhang C, Wu Z, Xu Y, Hu M, Dong Z. Distributed adaptive robust voltage/VAR control with network partition in active distribution networks. IEEE Transactions on Smart Grid. 2020; **11**:2245-2256. DOI: 10.1109/TSG.2019.2950120

[36] Landeros A, Koziel S, Abdel-Fattah MF. Distribution network reconfiguration using feasibility-preserving evolutionary optimization. Journal of Modern Power Systems and Clean Energy. 2019;7:589-598. DOI: 10.1017/s40565-018-0480-7

[37] Kanase AB, Gengaje SR. Reconfiguration of electrical distribution network using artificial neural network. International Journal of Advanced Research in Engineering and Technology (IJARET). 2018;**9**:119-128

[38] Oloulade A, Imano AM, Fifatin X, Vianou A, Tamadaho H, Badarou R. Multi-objective optimization of the safe operation on the electrical distribution system by placing D-FACTS and network reconfiguration. Journal of Power and Energy Engineering. 2019;7:94-113. DOI: 10.4236/jpee.2019.79007

[39] Pandey GP, Mali B, Shrestha A, Singh A, Jha AK. Performance enhancement of radial distribution system via network reconfiguration: A case study of urban city in Nepal. Journal of Renewable Energy, Electrical, and Computer Engineering. 2021;**1**:1-11. DOI: 10.29103/jreece.v1i1.3455

[40] Kahouli O, Alsaif H, Bouteraa Y, Ali NB, Chaabene M. Power system reconfiguration in distribution network for improving reliability using genetic algorithm and particle swarm optimization. Applied Sciences. 2021;**11**: 1-15. DOI: 10.3390/app11073092

[41] Zhou Q, Shirmohammadi D, Liu WH. Distribution feeder reconfiguration for service restoration and load balancing. IEEE Transactions on Power Systems. 1997;**12**(2):724-729. DOI: 10.1109/59.589664

[42] Zhou Q, Shirmohammadi D, Edwin Liu WH. Distribution feeder reconfiguration for operation cost reduction. IEEE Transactions on Power Systems. 1997;**12**:730-735. DOI: 10.1109/ 59.589665

[43] Ding F, Loparo KA. A simple heuristic method for smart distribution system reconfiguration. IEEE Energytech. 2012;**1**:6. DOI: 10.1109/ EnergyTech.2012.6304698

[44] Nuhanović A, Hivziefendić J, Hadžimehmedović A. Distribution network reconfiguration considering power losses and outages costs using genetic algorithm. Journal of Electrical Engineering. 2013;**64**:265-271. DOI: 10.2478/jee-2013-0039

[45] Atmaca E, Ersoz D. Testing ordinal optimization approach in power distribution system control. IEEE/PES Transmission and Distribution Conference and Exposition. 2008;**2008**: 1-5. DOI: 10.1109/TDC.2008.4517078

[46] The T. T, Ngoc DV, Anh NT. Distribution network reconfiguration for power loss reduction and voltage profile improvement using chaotic stochastic fractal search algorithm. Complexity. 2020;**2020**:1-15. DOI: 10.1155/2020/ 2353901

[47] Baran ME, Wu FF. Network reconfiguration in distribution systems for loss reduction and load balancing. IEEE Transactions on Power Delivery. 1989;**4**:1401-1407. DOI: 10.1109/ 61.25627

[48] Bakken D, Iniewski K. Smart Grids: Clouds, Communications, Open Source, and Automation, Chapter: Expert Systems Application for The Reconfiguration of Electric Distribution Systems. USA: CRC Press, Taylor and Francis Group; 2014. pp. 359-373

[49] Devabalaji KR, Imran AM, Yuvaraj T, Ravi K. Power loss minimization in radial distribution system, 2015 international conference on alternative energy in developing countries and emerging economies. Energy Procedia. 2015;**79**:917-923. DOI: 10.1016/j. egypro.2015.11.587

[50] Essa MB, Alnabi LA, Dhaher AK. Distribution power loss minimization via optimal sizing and placement of shunt capacitor and distributed generator with network reconfiguration. Telkomnika. 2015;**19**:759-766. DOI: 10.12928/ telkomnika.v19i3.15223

[51] Abdelaziz AY, Ali ES, Abd Elazim SM. Optimal sizing and locations of capacitors in radial distribution systems via flower pollination optimization algorithm and power loss index modified cultural algorithm. International Journal of Advanced Research in Engineering

and Technology (IJARET). 2016;**19**:610-618. DOI: 10.1016/j.jestch.2015.09.002

[52] Selim A, Kamel S, Jurado F. Power losses and energy cost minimization using shunt capacitors installation in distribution systems. In: 2019 10th International Renewable Energy Congress (IREC 2019). London: IEEE; 2019. pp. 1-6. DOI: 10.1109/IREC.2019.8754607

[53] Kamel S, Selim A, Jurado F, Yu J, Xie K, Wu T. Capacitor allocation in distribution systems using fuzzy loss sensitivity factor with sine cosine algorithm. IEEE PES Innovative Smart Grid Technologies Asia. 2019;**2019**:1276-1281. DOI: 10.1109/ISGT-Asia.2019.8881794

[54] William H. Kersting. Distribution System Modeling and Analysis. Second ed. Boca Raton, Florida, United States: CRC Press, Taylor & Francis Group; 2007. ISBN: 0-8493-5806-X

[55] Kleinberg MR, Miu K, Segal N, Lehmann H, Figura TR. A partitioning method for distributed capacitor control of electric power distribution systems. IEEETtransactions on Power Systems. 2014;**29**:637-644. DOI: 10.1109/TPWRS.2013.2281764

[56] Mori H, Komatsu Y. A parallel meta-heuristic method for distribution network loss reduction with network reconfiguration and capacitor control. Transmission & Distribution Conference & Exposition: Asia and Pacific. 2009; **2009**:1-4. DOI: 10.1109/TD-ASIA.2009.5356970

[57] Elkhatib ME, Shatshat RE, Salama MMA. Decentralized reactive power control for advanced distribution automation systems. IEEE Transactions on Smart Grid. 2012;**3**:1482-1490. DOI: 10.1109/TSG.2012.2197833

[58] Haldar V, Chakraborty N. Power loss minimization by optimal capacitor placement in radial distribution using modified cultural algorithm. International Transactions on Electrical Energy Systems. 2015;**25**:54-71. DOI: 10.1002/etep.1820

[59] Mahdavi M, Alhelou HH, Hatziargyriou ND, Jurado F. Reconfiguration of electric power distribution systems: Comprehensive review and classification. IEEE Access. 2021;**9**:118502-118527. DOI: 10.1109/ACCESS.2021.3107475

[60] Tinney WF, Hart CE. Power flow solution by Newton's method. IEEE Transactions On Power Apparatus And Systems. 1967;**86**:1449-1460. DOI: 10.1109/TPAS.1967.291823

[61] Peschon J, Piercy DS, Tinney WF, Tveit OJ. Sensitivity in power systems. IEEE Transactions On Power Apparatus And Systems. 1967;**86**:1687-1696. DOI: 10.1109/TPAS.1968.292130

[62] Robin J. Wilson, Introduction to Graph Theory. Fourth ed. Edinburgh Gate, Harlow, Essex England: Addison Weley Longman Limited; 1996

[63] Rong Z, Xiyuan P, Jinliang H, Xinfu S. Reconfiguration and capacitor placement for loss reduction of distribution system. In: 2002 IEEE Region 10 Conference on Computers, Communications, Control and Power Engineering. TENCOM'02. Proceedings. Vol. 3. London: IEEE; 2002. pp. 1945-1949

Chapter 4

Advances and Prospects in Distributed Generation Sources Digital Twins Design

Ivan Todorović and Ivana Isakov

Abstract

Power electronics devices are highly dynamic and nonlinear systems governed by complex control algorithms. In addition, power electronics converters must demonstrate high reliability and must be safe to operate, although in some applications they condition dangerously high power levels. Therefore, the development, evaluation and deployment of these systems traditionally had to be meticulously conducted, using specialized tools and approaches. With the proliferation of power electronics devices into the domain of power systems, in form of distributed generation sources, the mentioned tools and approaches had to evolve further. Consequently, a wide range of representations of the addressed systems within various digital platforms is nowadays at disposal of researchers and engineers. These representations, designated as digital twins, facilitate, and sometimes simply make possible, expeditious development and comprehensive evaluation of increasingly complex power electronics-based systems. This chapter catalogues the most important digital twin types, explicates their advantages and disadvantages and addresses their applicability. Hence, it could be regarded as a set of guidelines on how to choose appropriate digital twin type and digital twinning platform for some particular research and engineering problem. Also, details on how digital twins of distributed generation sources will be created and utilized in near future are provided.

Keywords: digital twin, hardware in the loop, real-time simulation, power system modeling, power hardware in the loop, cosimulation, distributed generation sources

1. Introduction

The power electronics (PE) devices consist of semiconductor switches, some of which are controllable (transistors, thyristors, etc.), while some are uncontrollable (diodes). Since semiconductor switches themselves are nonlinear elements (voltage–current dependency of the switches is nonlinear), the resulting power converter is also a highly nonlinear device. Additionally, several important PE topologies consist of many semiconductor switches. Consequently, addressing the *hardware* of PE devices can be a difficult task.

On the other hand, since controllable semiconductor switches necessitate control algorithms to govern their operation, and with a goal to secure different types of efficient electric energy conditioning, i.e., conversion—from AC to DC, from DC to AC, etc., the PE converters require complex *software* structures to be implemented on dedicated microcontrollers. Moreover, as for other automated systems, here also all subsystems in a closed feedback loop must operate properly or device destruction can take place.

On system level, PE devices must demonstrate high levels of reliability since they are integral part in many critical systems (e.g. systems for electric energy production, electric vehicles) or because they are embedded in mass-produced devices (e.g. consumer electronics). They are also not easy to repair, and their failure leads to prolonged non-operation periods. Furthermore, strict safety standards must be satisfied before product commercialization, so that the safety of humans is not jeopardized even during catastrophic failures. Similarly, the operation of PE devices must not compromise the proper operation of other devices in their vicinity. This is especially difficult if a converter is used in high-power applications. Also, electro-magnetic-interferences standards must be followed, which adds an additional layer of complexity to the PE design process.

Finally, the PE product development life cycle is affected by market dynamics and even PE devices must be designed and produced at an ever-increasing pace. In this context, seemingly the most meaningful trajectory from idea to an operational device is to synthesize a sketch of a device (hardware and software aspects) and immediately build a prototype. Every young engineer has tried this approach and has failed.

For these reasons, conceptualization, design, implementation, integration, and system verification of these multidomain devices must be conducted both systematically and rapidly [1]. While others depend mostly on the skill and experience of the developers and cannot be easily changed and improved, the system verification depends heavily on which tools are used and consequently can be done in a more or less optimized fashion.

The system verification should not be done on a prototype at the beginning of the product development process, as indicated previously, but rather at its end. Before that, verification of certain subsystems and features has to be conducted within safe and flexible environments. This is true for PE devices in general, but it is especially so for PE systems that are part of distributed generation sources (DGSs). In these systems, mistakes made during development processes are exceptionally expensive, time-consuming, and dangerous for equipment operators.

PE solutions' verification tools are numerous, but most of them are used to build a digital representation of the addressed system within a certain type of digital computer. Such representations are called either digital twins or simulation models, since they are built with simulation software. This migration of PE devices into the digital domain and usage of digital twins, although being an additional step, actually accelerate the development process. That is because conducting tests within software environments is completely safe, enables test automation, accelerates design errors identification and correction, allows for certain parts of the design or the whole design to be shared more easily among the researchers and engineers (facilitates cooperation), gives freedom to developers to risk more and try novel solutions, not worrying about making mistakes, etc. Hence, digital twinning has been an integral part of the development process of practically all PE device types, excluding only the simplest ones. Moreover, digital twinning has been in use essentially ever since the rise of personal computers (PCs). Accordingly, there are many types of tools for digital twinning that have been developed over the years, each solving a specific set of design problems.

The following three sections give details on the three most important approaches for digital twin's generation and usage. The advantages and disadvantages of these approaches are given and when it is meaningful to use certain digital twinning tool is described. The digital twinning tools based on personal computers (offline simulation tools) are addressed in the second section. Those based on dedicated digital platforms (real-time simulation tools) are analyzed in the third section, while those considered emerging approaches are considered in the fourth section. The last section brings concluding remarks.

The following two statements should be noted. DGSs will be of particular interest in the remainder of the text because of their complexity and wide social, economic, and ecological importance. Still, most of the concepts and approaches provided in the text can be applied during the digital twinning processes of other power electronics devices. Also, models are never verbatim digital representations of physical systems, and some phenomena are always neglected or abstracted, but it is up to a developer to define how detailed the models will be.

2. Offline simulation platforms for DGSs digital twinning

Ever since the inception of PE as a major electrical engineering field, it became obvious that it would be rather arduous to develop PE devices and examine their behavior analytically, particularly their transient behavior. The first platform that enabled accelerated development, especially the validation stage, was personal computers (PCs). Many simulation tools that could be used to build digital representations of DGSs and PE devices, in general, were developed during the eighties, quickly after personal computers became widely available. The fact that many of these tools are still in use today is a testimonial of their usefulness and efficacy.

Initially, to build a digital twin and use it as an investigative tool using the offline simulations approach, the researchers and engineers needed only a PC and a single software tool (simulation tool or engine). Nowadays, offline simulations can be conducted in a slightly more complicated way, but the PC, i.e., general-purpose device, is still the main platform.

Depending on what part of DGS is of particular interest and which operational domain of the device should be put under scrutiny, four main types of offline simulations can be used today:

- Model in the loop (MIL);
- Software in the loop (SIL);
- Processor in the loop (PIL);
- Controller model in the loop (CMIL).

Several simulation software can be used in either of these approaches, but the tool itself must be optimized and additionally set up, i.e., cannot be used interchangeably out of the box.

The noun before "in the loop" in the approaches' names generally (excluding MIL) designates which part of the system is modeled in more detail or which part of the system is a real device, i.e., device under test (DUT).

2.1 Model in the loop

A digital twin of the addressed system always consists of digital representations of hardware (power stage) and software (control scheme) subsystems, correspondingly with real PE device. The power stage, or more precisely sensors' representations, provides the control scheme with the information about controlled variables, and the control scheme generates control signals for semiconductor and other active devices in the power stage. In the case of MIL, both power stage and control structures, governing the power stage, are implemented within the simulation tool using the blocks and elements that are native to the simulation tool. Hence, the digital twin consists only of the simulation model and the model is in the loop (is analyzed). The MIL approach corresponds to a traditional offline simulation. This environment consists of a PC with the installed simulation software, as indicated in **Figure 1**.

Since the power stage can be comprised of different PE devices, machines, power systems' parts, and electrochemical elements, the simulation tool must either have rich library elements or enable the user to develop his own parts using available blocks. Either way, a graphical user interface is used in modern simulation tools. This graphical representation is then compiled and run within the same software environment. The difference between the execution time, i.e., how much time is necessary for compilation and running of the model, and simulation time depends on model complexity and the PC performances. Since DGSs are complex, the execution time is significantly longer than simulation time (at least one order of magnitude). Once the execution is done, the developer can inspect the model's behavior and introduce changes if necessary.

Different simulation tools that enable the MIL approach focus on different aspects of DGSs. It is true that different software packages can be used to focus on more than one aspect of the DGSs, but they are usually optimized for one or maybe two layers of abstractions pertinent to DGSs. If semiconductor driver circuits, parasitic phenomena, and generally detailed simulation of solely PE devices (not the whole DGS) are necessary, the user can use several free (LTspice [2], Xyce [3]), and licensed (PSIM [4], PSpice [5], etc.) software tools. Alternatively, if the behavior of the complete DGS should be analyzed, the user again can choose between free (Typhoon HIL's VHIL [6]) and licensed (MATLAB/Simulink [7], PSIM, PLECS [8]) packages. Lastly, if the DGSs

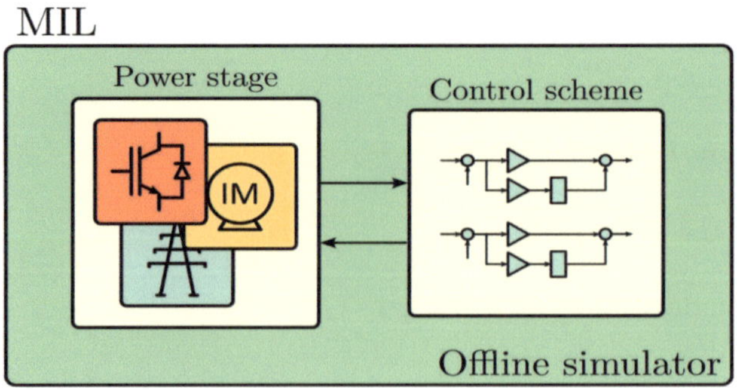

Figure 1.
Offline simulation approach—Model in the loop.

are to be integrated with the large power systems and jointly analyzed, users can employ several mature tools (PowerFactory [9], PSCAD [10], EMTP [11], etc.).

MIL approach was the first to be utilized to create DGSs digital twins and that pertinent simulation tools and have been in use for decades. From this stems the main advantage of MIL. MIL simulators have rich libraries, many readily available examples and developed user's communities. Consequently, it is both easy to start using these tools and to start new projects. Also, there are many plugins and add-ons that expand the software functionalities, albeit these additional features are not always free. Furthermore, MIL software is reliable. In addition, MIL in principle allows arbitrarily complex and detailed models to be implemented, at the price of slow model execution.

On the other hand, the MIL assumes utilization of PC—a general-purpose device that certainly is not optimized for running simulations software. In case when DGSs are analyzed, this results in long, sometimes impractical, execution times, especially if DGSs are integrated with a certain power system. Moreover, the MIL approach in many aspects gives crude information on how the real system is going to behave. If semiconductor phenomena are included in the model, for example, it becomes impractical to analyze wide power system dynamics. Alternatively, if large systems are analyzed, transient phenomena in PE devices are not captured properly. The most problematic aspect of MIL usage in this context is the estimation of how DGS control software will behave once deployed on the real device, i.e., microcontroller.

Consequently, the MIL approach is usually employed as the first stage in the DGS design verification process or when a singular feature is to be tested.

2.2 Software in the loop

SIL approach is quite similar to the MIL approach. It is also based on a PC, and it consists of one software package, usually the same as in the case of MIL. The power stage can be the same in both cases. The same information is exchanged between the power stage and control scheme. Hence, it brings similar features, advantages, and disadvantages. Still, it has one crucial difference. As **Figure 2** suggests, the software is not implemented using the same elements that are used for building the hardware part of the digital twin.

The software is realized utilizing certain embedded language (usually C or C++)—the same language that will be used to program the real microcontroller. Hence, SIL, besides providing a functional test of the control algorithm, enables the developer to gain insight into how the code is structured and organized, that is, it includes

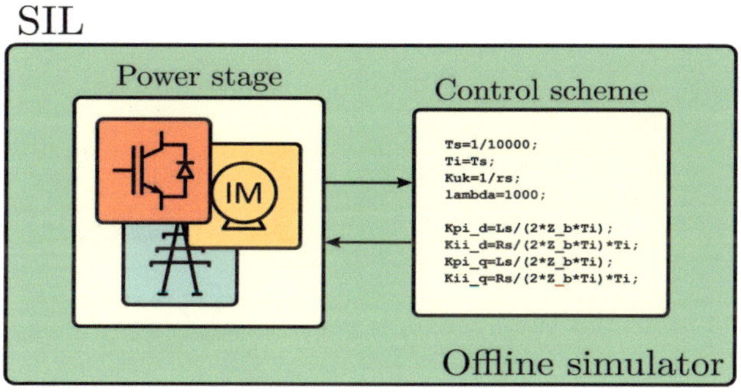

Figure 2.
Offline simulation approach—Software in the loop.

semantic and syntactic checkups. Nowadays, embedded code compilers that must be invoked implicitly during model execution are available for most popular simulation packages [12].

2.3 Processor in the loop

To investigate further control code features and behavior, the code developed in the simulation tool is not placed in the PC's memory, but it is downloaded to a real controller that shall be used in the final design. This approach is called "processor in the loop" (PIL) since the processor of the real controller is used. The embedded code compilers are again called during model execution. There has to be a communication line established between the controller and offline simulator (PC), so that relevant information between the power stage and control structures is exchanged, as in the previous two cases. The power stage developed in the simulator is again the same, and the model is executed usually significantly slower than in real time, because of the simulated power stage complexity. This implies that the control code execution on the microcontroller has to be slowed down (otherwise it could be run in real time) so that it is synchronized with the model execution.

In addition to advantages and disadvantages inherited from the SIL approach, the PIL approach allows control card memory and processor-time utilization to be examined. It should be noted that not all control card hardware peripherals are utilized and tested—most notably, pulse width modulation peripheral and analog-to-digital conversion unit are not put under test. Additionally, PIL has a bottleneck in the number of DGSs that can be modeled simultaneously. Since one communication channel can be formed between the offline simulator and the controller board, the real controller can drive only one DGS. Also, the toolboxes necessary to utilize PIL are not available for many controllers, but only for flagship models (**Figure 3**) [13].

2.4 Controller model in the loop

Controlled code execution and controller board's resources utilization can be studied in even more detail. For that, comprehensive behavioral models of microcontrollers can be developed. Although the idea is not new, only recently several microcontrollers' vendors have developed these models and made them available for usage on PC [14]. It should be emphasized that the controller model is usually executed on software different from simulation software, and these two software packages must communicate

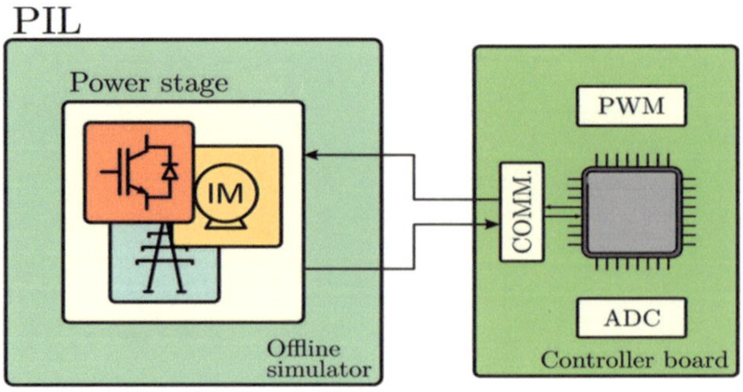

Figure 3.
Offline simulation approach—Processor in the loop.

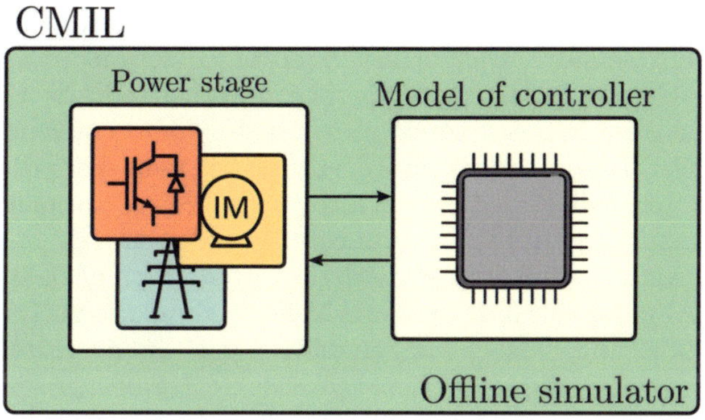

Figure 4.
Offline simulation approach—Controller model in the loop.

and exchange relevant information between the power stage (simulation tool) and control scheme (software dedicated to the simulation of the controller behavioral model). Hence, the controller model in the loop approach assumes usage of only PC (without any external hardware), as MIL and SIL (**Figure 4**).

The feature of the controller model in the loop approach is that developers can see exactly how the controller and implemented control code are going to behave once deployed on the real controller—how the memory is going to be utilized, will there be computational overload, how the latency will affect the system operation, etc. Also, the code embedded in the simulated controller model can be immediately deployed on the real controller, without any changes.

This main novelty of this approach, in the context of DGSs, is its main drawback. CMIL puts focus on the controller and control code implementation. Moreover, as for PIL, the setup is essentially limited to one controller and one DGS.

3. Real-time simulation platforms for DGSs digital twinning

The proliferation of renewable energy sources and DGSs into the traditional power systems during the beginning of the twentieth century brought new development challenges. This amalgamation of the two fairly different domains of electrical engineering challenged the existing development and validation tools. PE devices and DGSs necessitate small simulation time steps (high simulation resolution) to capture properly dynamic phenomena associated with these systems. The power systems are by themselves large and complex and developers used offline simulation tools that tackled this problem by increasing simulation time steps—which was acceptable since all relevant phenomena in traditional power systems were characterized by long time constants. This translated to the possibility of setting essentially arbitrarily long simulation time when simulating power systems. Hence, to address properly these emerging systems, both small simulation time steps and long simulation times were necessary. This predicament was exacerbated further by the fact that tests had to be repeatedly conducted (considering different operating points, networks configurations, control functions, etc.). Consequently, the usage of traditional offline simulation tools for modeling power systems integrated with DGSs became impractical.

The main impediment was the platforms used to conduct traditional offline simulations—PCs. To enable more efficient and practical validation of DGSs integrated with power systems, new, dedicated, digital platforms had to be developed.

The first efforts went in the direction of paralleling many conventional processors [15]. Although this gave certain improvements (systems that are more complex indeed could have been addressed), the scalability, firmware complexity, and maintenance of such systems were impractical for commercial products.

The paradigmatic change happened with the introduction of field-programmable gate arrays (FPGA) platforms in the domain of DGSs simulations. They provide parallel computations and consequently enable a significant decrease in models' time execution.

Actually, not only that complex models' execution was accelerated, it became possible to run models in real time. Besides accelerating validation procedure, the usage of FPGA-based platforms enabled interaction of real-time executed models and real devices, in which cases the real device was a device under test (DUT), and the model was used to mimic sophisticated operating conditions, often found in power systems rich with DGSs.

Consequently, two real-time simulation approaches emerged:

- Controller hardware in the loop (C-HIL);

- Power hardware in the loop (P-HIL).

Both approaches are based on real-time simulators (RTSs). RTSs are interfaced with either microcontroller running designed control schemes or with certain high power devices (e.g., inverter).

3.1 Controller hardware in the loop

The most comprehensive microcontroller and control schemes operation analysis can be conducted using the C-HIL approach, especially if distribution networks or microgrids with many DGSs are considered. The real controllers, with control code that will be deployed in the field to drive DGSs, are used, and their interaction with the rest of the system can be examined considering phenomena pertinent to networks proliferated with DGSs. How precise are the test results depends only on how precise are the power stage models. It should be emphasized that the models' complexity, precision, and level of details are as good as in the case of offline simulator approaches, if not better. The power stage complexity does influence the simulation time steps, but modern RTS enables sub-1 μs time steps, even for complex power stages, which is sufficiently small for DGSs applications. Alternatively, C-HIL can be used to test the interaction of one converter and its surrounding subsystems (e.g., grid-connected converter with stiff grid, electric vehicle, etc.). Finally, this environment can be used to test the operation of the singular converter, such as an inverter in machine-driven applications. In other words, the same C-HIL platform can be used to analyze PE systems of almost arbitrary complexity.

Still, it should be said that the C-HIL is usually not intended to be used for testing semiconductor-devices-related phenomena, although there are no technical obstacles [16]. Actually, semiconductor devices are oftentimes modeled as ideal switches. The semiconductor-devices-related phenomena can be more meaningfully addressed using the MIL approach, utilizing free MIL software.

The C-HIL environment consists of real-time simulators, interface boards, and controllers. The outline of the C-HIL platform is given in **Figure 5a**. The RTSs are often denoted as emulators, as they emulate the behavior of certain power stages. The complexity of the power stage can range between singular and simple DC-DC converter (e.g., buck converter) to a distribution network of a small town with many DGSs included.

The power stages are defined on PC, using dedicated software, compiled and downloaded on emulators. Although emulators themselves are rather complex digital systems, their complexity and complexity of power stages are usually hidden from the user behind IO boards.

The interface board (IB in **Figure 5**) is used to adjust analog and digital signals exchanged between the emulator and the controller. The emulator sends analog signals that should be analogous to the signals that would be generated by the real sensors in the real power stage (currents, voltages, machine speed, etc.). These signals are adjusted using ordinary operational amplifiers circuits (AMP in **Figure 5a**). Then they are forwarded to the controller's analog inputs and analog-to-digital conversion units. This information about relevant power stage variables is used in the controller to synthesize new control and gating signals. Accordingly, the controller generates

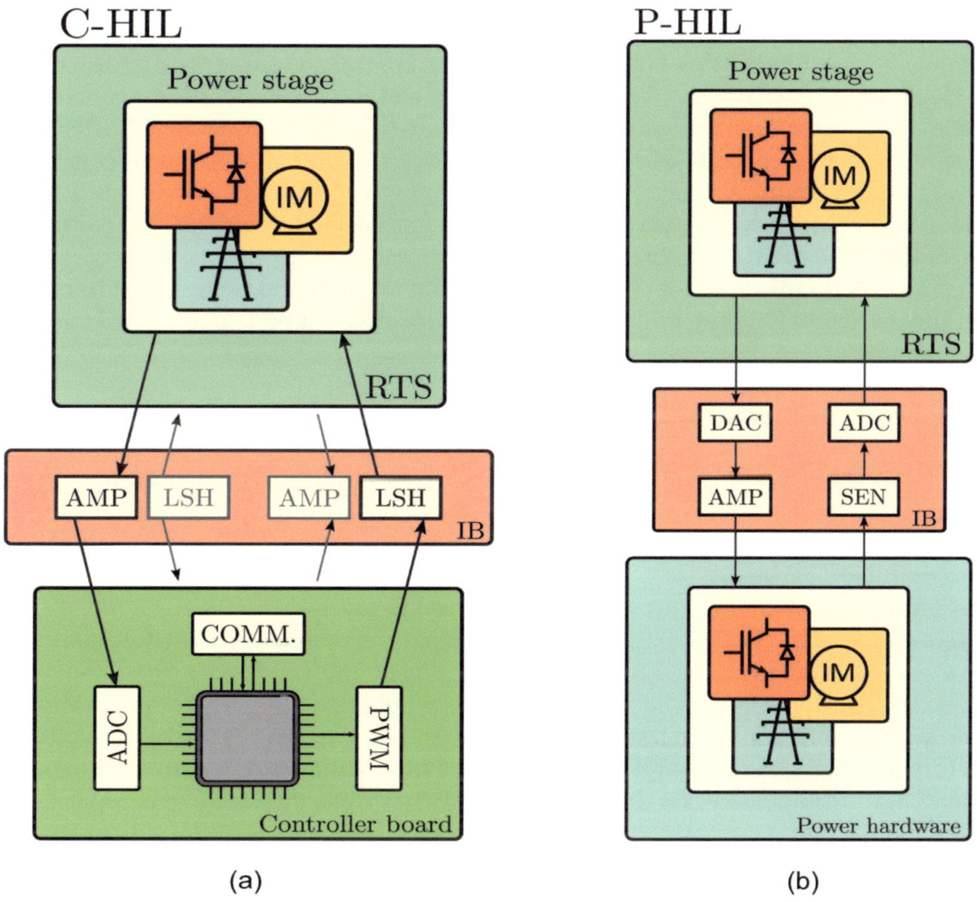

Figure 5.
Real-time simulation approaches—Controller model in the loop (a) and power hardware in the loop (b).

digital, gating, signals for the converters' switches, found in the power stage. These signals are also adjusted using level shifters (LSH in **Figure 5a**). Then these signals are sent to the RTS's digital inputs. There, they result in turning on and off certain switches, resulting in circuit reconfiguration and adequate change in controlled variables. Hence, the feedback loop is closed. Analog and digital signals can be exchanged in the reversed fashion also, but those signal pathways are of secondary importance in most of the DGSs-related applications. Since operational amplifiers and level shifter circuits are simple, the whole interface boards are simple electronics circuits and can be quickly developed for any controller board.

Controller boards used in C-HIL, which is a DUT in this setup, can be arbitrarily chosen with the only limitation that they must have sufficient resources to run the control code that should secure necessary DGS's features.

The C-HIL setups enable the creation of comprehensive DGS' digital twins and high-fidelity testing procedures. The users are provided with reliable control code behavior testing results, without compromising either the equipment nor personnel. The fact that no high power devices are used makes this environment perfectly safe and equipment management is done without procedural hurdles. Since real-time tests execution is secured, the tests execution is significantly accelerated, in comparison to offline simulation approaches. Actually, if it was not for controller boards, the test could be done even faster than in real-time—the RTSs are sufficiently quick, but are "slowed down" so that they can be meaningfully interfaced with controller boards. Moreover, considering that both power stage and control code are run in a form of code on a specific platform (RTS for power stage and controller boards for control code), test automation is possible [17]. Complete testing procedures (with different operating points, control codes, network parameters, etc.) can be executed consecutively and with minimal personnel effort by running simple scripts, usually written using python. Finally, nowadays, the RTSs are affordable, and their price is comparable with MIL or SIL software licenses.

The only disadvantage of C-HIL usage to build DGSs digital twins and run tests on them is that it is not possible to examine the operation of high-power hardware components.

3.2 Power hardware in the loop

P-HIL setup was proposed to address the shortcoming of the C-HIL approach—as an environment in which certain high-power devices can be tested in conjunction with an adequate digital twin. DUT can be anything from a simple protection device to a multilevel converter.

These setups could be meaningfully deployed in several situations. Firstly, in complex PE and DGSs systems, the system's parts are developed at different instances. Hence, to manage efficiently development and verification time, the functional subsystems can be tested before the whole setup is ready for integration and verification. Similarly, it is generally advisable to test parts of a complex system before operating on the complete system, since this facilitates design mistakes identification and decreases the possibility of catastrophic failures occurring during final tests. In these cases, the P-HIL can mimic the behavior of the still-to-be-implemented part of the addressed system. Secondly, when addressing high-power devices, the P-HIL environment is advantageous since it can be safer for verification than standard full-power prototypes—prompt termination of dangerous and faulty tests is inherently easier in a P-HIL setup. Thirdly, if the behavior of DGS's hardware should be investigated against complex network conditions (e.g. protection device of a grid-connected converter in microgrids), the P-HIL is irreplaceable since tests on real networks would be either impossible or expensive and dangerous for both equipment and personnel.

The P-HIL platforms, depicted in **Figure 5b**, are conceptually similar to C-HIL counterparts, but have some specificities also. The RTS part of the setup can be the same emulator unit as in the C-HIL setup, but the model realized in the RTS is generally either passive network or active network with simplified (averaged) models of DGSs, since there are no controller boards controlling the DGSs' operation.

The interface cards are in this case significantly different. This is a consequence of the fact that DUT is a high-power device. Hence, to connect and exchange relevant data between the RTS and DUT, the interface board must contain, besides ordinary operational amplifiers circuits and digital-to-analog conversion units (DAC in **Figure 5b**), high-power amplification circuits (AMP in **Figure 5b**). AMP in **Figure 5a** and AMP in **Figure 5b** are not the same circuits. High-power amplification circuits are rather complex devices and can be based on switching amplifiers, linear amplifiers, synchronous machines, or multilevel converters [18]. It is evident that the amplification circuits can be as complex as DUT itself, and they are only a part of the interface board. Moreover, high-power amplification units inherently introduce signal propagation bandwidth limitation and latency. This can compromise experiments' precision and reliability and can cause stability issues. Mitigation of these issues is a topic of ongoing investigations [19]. The interface board, besides generating high power variables (currents, voltages, and mechanical variables), must adjust analog variables coming from DUT toward the RTS. For this, sensors and analog-to-digital conversion circuits must be implemented. This complicates the interface board further.

The power hardware, i.e., DUT, can be any piece of hardware pertaining to PE or DGS device whose operational characteristic shall be examined.

Consequently, the P-HIL offers a comprehensive analysis of how certain real pieces of hardware will behave once deployed in the field. It is a relatively safe environment, but since it does consist of high-power devices, the flexibility and ease of use of P-HIL are decreased in comparison to C-HIL setups. Moreover, interface boards' complexity limits to some extent the commercial attractiveness of P-HIL setups. Finally, although P-HIL setups are based on emulators and are easily reconfigurable, testing automation cannot be fully realized, and some manual interventions must be made during different experiments being executed consecutively.

4. Emerging platforms for DGSs digital twinning

The C-HIL and P-HIL environments were until recently sufficient to meet all the digital twinning needs of engineers and researchers in the domain of DGSs. Several niches, but important, research and public sphere incentives have resulted in the development of advanced platforms for digital twinning. They are all based on C-HIL and P-HIL devices, but are conceptually somewhat different and bring several new features.

Currently, the following types of advanced digital twinning platforms are developed:

- Cosimulation platforms;
- Cloud emulation and simulation platforms.

4.1 Cosimulation platforms

The first cosimulation platform version is intended for the emulation of large power systems. MC-HILs consists of multiple interconnected hardware in the loop, i.e., RTS units. The RTS units usually are from the same vendor, but they can be from different suppliers, so that the advantages of different units can be put into practice. Moreover, the microcontroller units can be different if serving different purposes. For example, if microgrids with many DGSs are addressed and if both low-level and high-level control structures are implemented, such setup would be rather convenient. This cosimulation type is depicted in **Figure 6a**. It essentially brings the same advantages and disadvantages that the C-HIL approach brings. The only differences stem from the fact that large systems can be addressed and that the setup itself is more complex and consequently somewhat harder to manage. Still, since there are no high-power devices, MC-HILs setups are versatile and easily reconfigurable.

Modern RTSs consist of both the FPGA platform (used for power stage emulation) and digital signal processing, i.e., microcontroller-like, part. The latter can be used to implement different control functions, ranging from simple network reconfiguration functions to low-level, time-critical, PE-related control structures. Hence, the control algorithms that were.

previously implemented on the dedicated controller cards are now implemented on the same unit as emulated power stage—no external controller cards are necessary. The HILs setup is shown in **Figure 6b**. A mixture of MC-HILs and HILs setups can be used, also (control structures can be implemented both on dedicated controllers and the emulators).

The HILs configuration has a disadvantage in that the control code is not run on the separated controllers, and consequently the control code execution and the controller cannot be directly validated. Still, if this is not of primary interest, but rather an acceleration of control structures development and validation, particularly in the

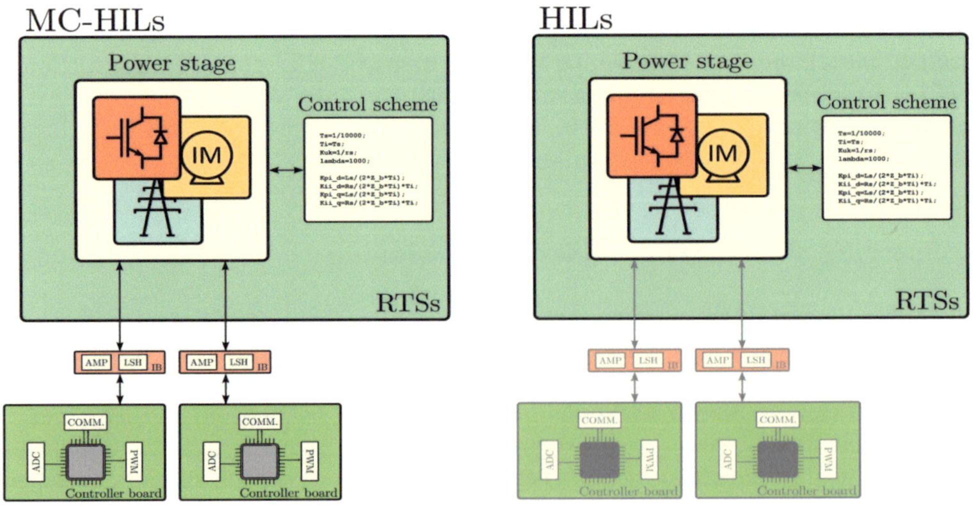

(a) (b)

Figure 6.
Cosimulation platforms—MC-HILs (a), HILs (b).

context of large power systems, ease of use and environment's management simplicity render HILs a most meaningful platform in this regard.

The third cosimulation platform is a composite of C-HIL and P-HIL approaches. The representation of CP-HILs setups can be found in **Figure 7**. Such a framework could find application in cases when grid-connected inverter's hardware and software should be tested against realistic active grid operating conditions, for example. To test inverters' parallel operation, several inverters can be emulated on the RTS and driven by the dedicated controllers, while DUT would be connected to the same virtual point of common coupling. Such tests are as close as possible to tests executed on real, complete setup, whereas a test in real operating conditions would be unacceptably expensive, dangerous, and impractical, especially during the validation phase and particularly at high-power levels.

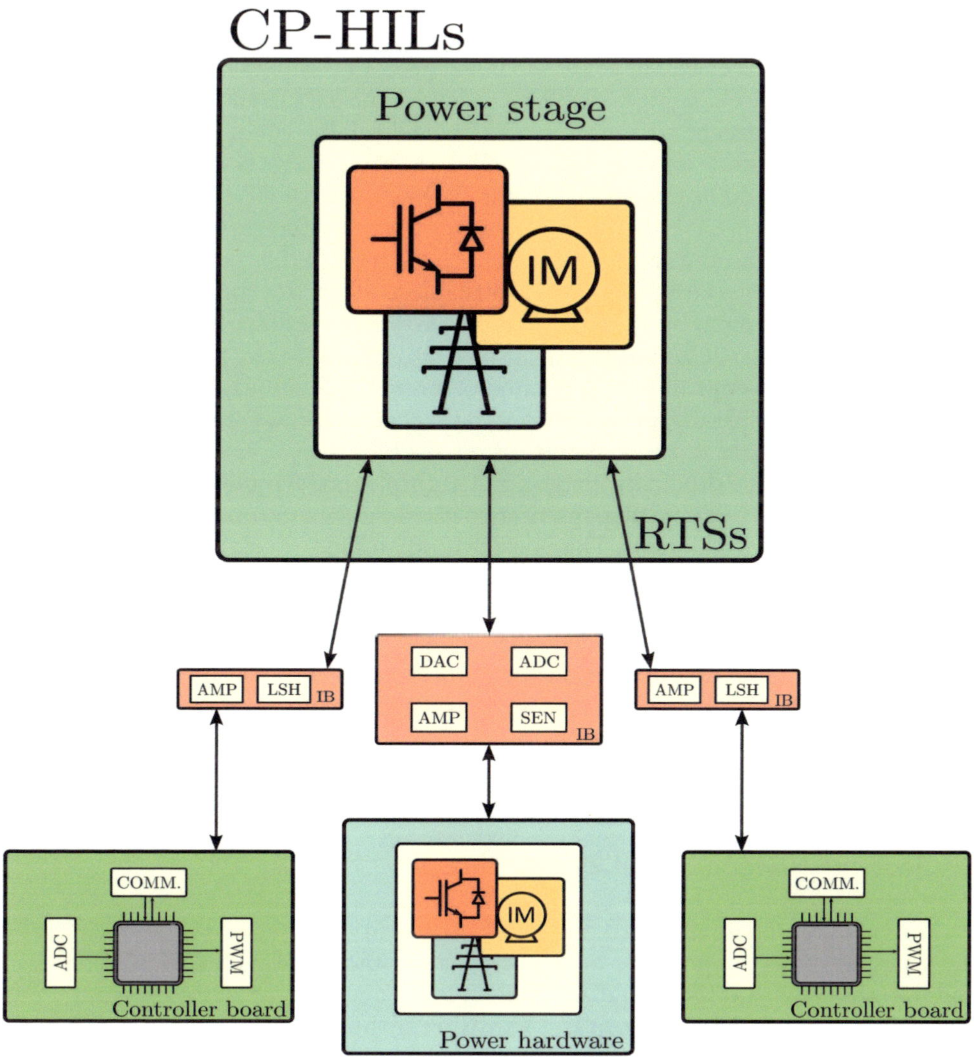

Figure 7.
Cosimulation platforms—CP-HILs.

Since CP-HILs' environment consists of all parts and devices found both in C-HIL and P-HIL paradigms, CP-HILs inherit all advantages and disadvantages of those two approaches, but are also much more complicated to manage, especially if several different RTSs and controllers are used. Furthermore, CP-HILs setups tend to be expensive.

The fourth cosimulation concept represents an aggregation of geographically dispersed C-HIL, P-HIL, and CP-HILs systems. The setups are connected over a dedicated Internet connection, and the emulation data are exchanged over this connection so that the cumulative platforms' emulation data can be synthesized. Expectedly, the necessity of transferring data over large distances results in significant data latency. Hence, if the whole model is to be executed at the same time step, the step time must be significantly larger in comparison to one in the C-HIL approach. This problem can be somewhat mitigated if the model is partitioned in accordance with the geographical arrangement and different parts are executed at different time steps (or generally in a nonsynchronized manner). Then the latency would affect, i.e., artificially increase "data sampling" period of only those variables that are exchanged between the setups—the rest of the model's parts could be executed at a much faster pace, in compliance with local setups capabilities. If slowly changing variables are chosen, the data latency does not affect significantly model execution and the reliability and validity of the results.

The first meaningful situation for Geo-CP-HILs setup usage is when the most complicated models should be emulated and one C-HIL, P-HIL or CP-HILs setup would not suffice. Next, Geo-CP-HILs platforms are used when interested parties want to conduct joint tests, but the data privacy must be secured. In these setups, the parts of the model are not necessarily known and available to all parties conducting the tests. Only the exchanged data, found at the model's parts intersections, are available to more than one test participant (this can also be limited to two test participants). This is the case when power grids to be emulated cover multiple countries, or otherwise administratively, politically, or socially separated territories and entities that want to preserve the security of sensitive power system data. Moreover, Geo-CP-HILs are used when complex models should be merged and multiple model implementations are to be evaded. Only one test's participant implements a model only once and afterward shares it with other participants. This is particularly important if the participants are using different platforms—the porting of the model then would be particularly difficult. Establishing Geo-CP-HILs setups can be financially meaningful when the setup usage time can be sold to other parties, as in time-sharing financial constructs.

There evidently are applications in which these frameworks would be applicable. Actually, Geo-CP-HILs indeed are the universal digital twinning platform, but considering **Figure 8**, it becomes obvious that Geo-CP-HILs systems are extremely complex, hard to manage, and difficult to constitute. Consequently, only a handful of such setups can be found worldwide [20, 21].

4.2 Cloud emulation and simulation platforms

At the beginning of the third decade of the twenty-first century, the technical and technological problems are not the biggest obstacles for wide DGSs' adoption in residential and industrial areas, but also generally in distribution and transmission networks. Software and hardware technologies pertinent to DGSs, excluding those applied to islanded microgrids, are mature and only incremental advancements are necessary. One of the biggest hurdles impeding further proliferation and "crossing the chasm" of the DGSs is related to how the general public

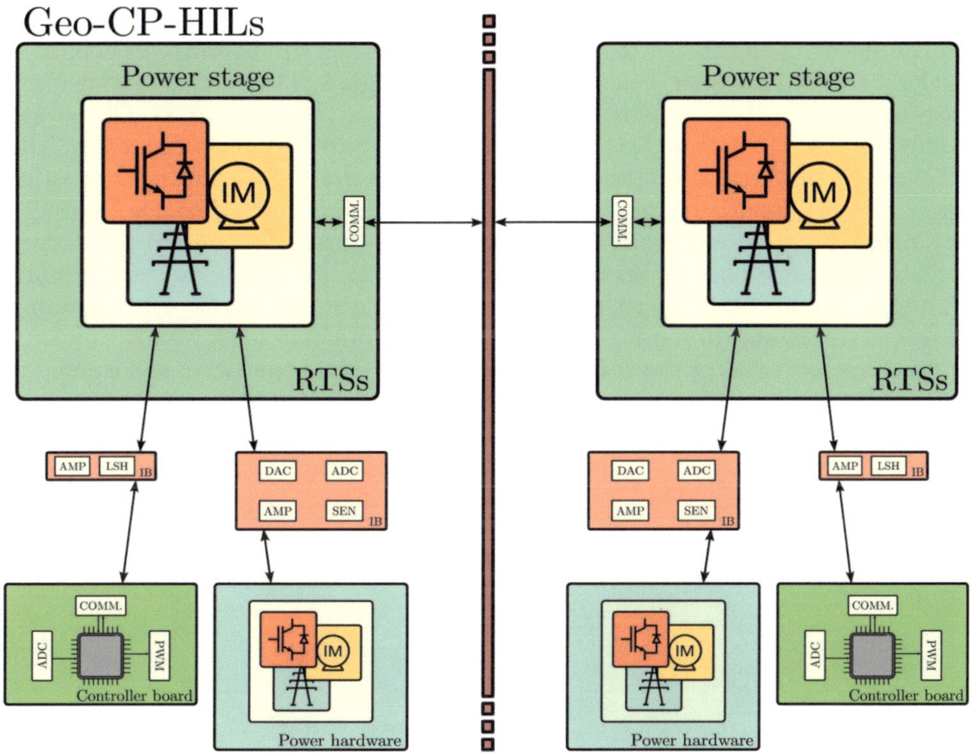

Figure 8.
Cosimulation platforms—Geo-CP-HILs.

perceives traditional grids and DGSs. The networks as they are operate reliably and efficiently and average resident and industrial entities are accustomed to how networks function. Moreover, DGSs are not technologies about which the average citizen is particularly knowledgeable. Hence, it may seem questionable why changing something that works well from the user standpoint. Especially since it should be replaced with a complex paradigm under which expensive renewable resources are used, energy storage systems are desirable, where third parties manage new assets so that they are (more) economically feasible and so that the system operates reliably.

The users must be provided with the best possible and palpable proof that the investment will be profitable and that the system will be reliable and safe. In order words, the investment in DGSs must be derisked.

For such tasks, the C-HIL platforms seem reasonable and promising. Using the C-HIL paradigm, comprehensive digital twins of the networks that should be upgraded with the DGSs can be made. Moreover, all control layers that will be deployed in a real network can be executed on a C-HIL setup. Consequently, using historical data (for solar irradiation, wind speed, load profiles, etc.) and other pertinent data within C-HIL digital twins, it becomes possible to quickly execute all kinds of technoeconomic studies. Such studies are the next best thing to data obtained from the real operating power system.

Still, the C-HIL setups are not as affordable so that they can be bought anytime certain party necessitates a power system case study. Moreover, they are intended to be operated by professionals.

Therefore, cloud emulation and cloud simulation platforms are being developed [22]. The primary objective is to "hide" the emulation and simulation tools behind the cloud services and enable electrical engineers, engineers from other engineering trades and non-engineering personnel to conduct tests. These platforms will provide automatic model generation (in accordance with data provided by users), compilation and execution of the models. Hence, the user will just have to provide the data defining the model, network topology, DGS installed power, historical data, etc., and will get the testing reports automatically. Consequently, cloud emulation and simulation platforms will act as black boxes for C-HIL or HILs setups from the user standpoint. It is important to note that third-party services, such as financial and energy-trading platforms, will be able to access the emulation and simulation tools.

There are two types of cloud services developed—cloud emulation and cloud simulation services.

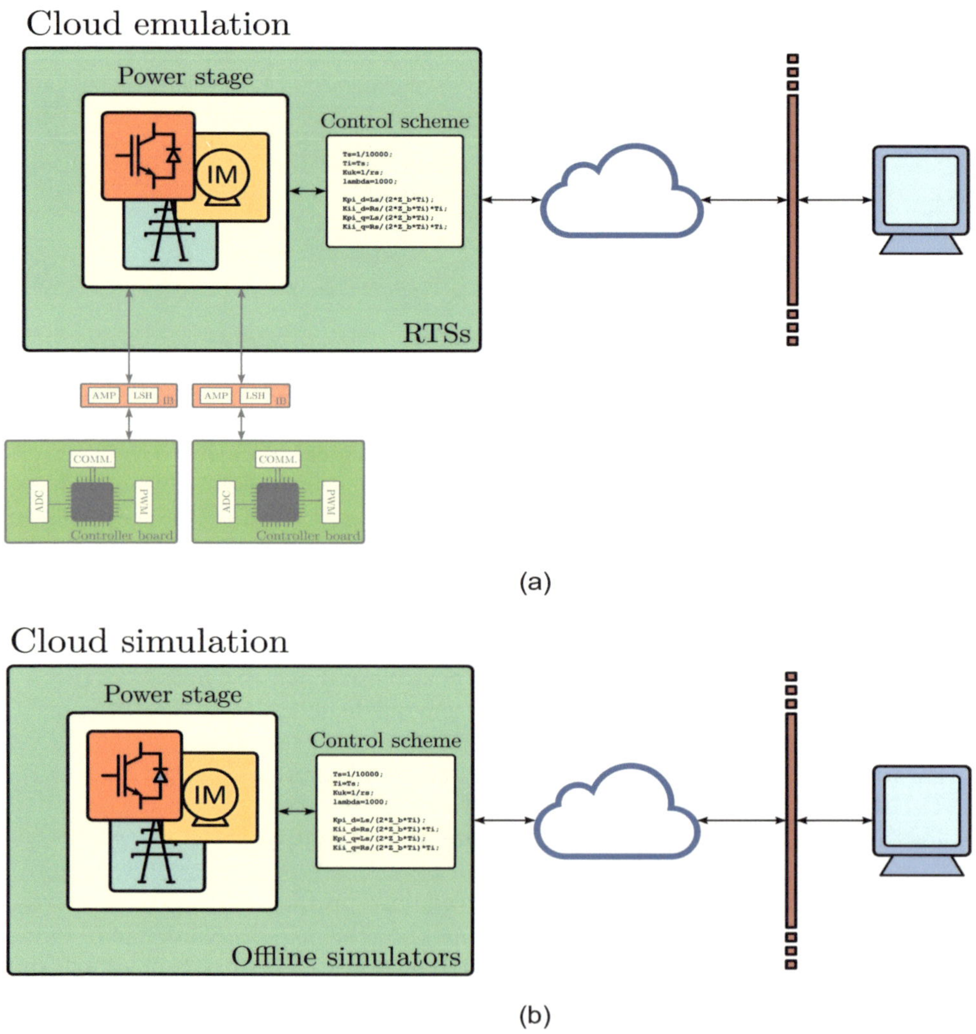

Figure 9.
Cloud emulation (a) and cloud simulation (b) platforms.

The cloud emulation platforms are based on RTSs, i.e., C-HIL or HILs setups, and inherit their performance characteristics. The outline of this paradigm is shown in **Figure 9a**. Once again, an external controller can be used, as in C-HIL, but certain RTSs have sufficiently large resources to run in real-time control code also and thus eliminate the need for dedicated controllers. In the domain of cloud services, it is irrelevant what will be the control code implementation platform. This enables further improvement. Namely, since both power stage and control code are run (more precisely designed, compiled, and run) on the same platform, there are no technical obstacles why both parts of the digital twin would not be run even faster than real time. The state-of-the-art RTSs enable up to two times faster than real-time models execution for medium complexity models, such as a network with 10 nodes and an equal number of DGSs. For more complex and larger models, the acceleration is naturally smaller, and vice versa. It should be emphasized that the RTSs, and hence cloud emulation services, can be used to examine even transient phenomena (phenomena with time constants smaller than 1 ms).

Alternatively, if transient phenomena are not of importance, but steady-state and averaged network behavior, the digital twin can be created using simplified network and DGSs elements. This is suitable for implementation using simulation tools on PC or server devices. This organization is designated as cloud simulation platform. It is conceptually depicted in **Figure 9b**. Since the digital twin is simpler than the digital twin implemented on emulation platforms, the execution time is significantly shorter. Depending on the network size, the execution time can be more than a thousand times shorter than the real time. This enables simulation of one year of network's operation within one day or less.

Once completely functional, cloud emulation and simulation platforms will become an irreplaceable tool for emerging power systems operation examination and derisking.

5. Conclusion

The power electronics, distributed generation sources, and power systems research and development processes are layered and complex, and there are a plethora of different tools that can be used to help engineers and researchers carry out their activities. Consequently, even the terminology can be confusing at times, and it is not always transparent which tools should be used during different development and validation phases for different applications.

This chapter gives an overview of platforms and paradigms that can be employed to create and utilize digital twins pertaining to mentioned systems. How digital twinning platforms are constituted and what are their main advantages and disadvantages is explicated. Also, details regarding the area of applicability are provided.

The chapter addresses traditional digital twinning platforms (model in the loop, software in the loop, etc.), real-time platforms (controller hardware in the loop and power hardware in the loop), and emerging platforms for simulation and emulation of advanced power systems proliferated with distributed generation sources.

Funding

The research was funded by "Innovative scientific and artistic research from the FTS activity domain 451-03-9/2021-14/200156" project, financed by Serbian Ministry of Education, Science and Technological Development.

The authors would like to thank the European Commission and the partners of the European Horizon 2020 “CREATORS—CREATing cOmmunity eneRgy Systems” (https://www.creators4you.energy/) for their help and support. The CREATORS project has received funding from the European Union’s Horizon 2020 research and innovation programme under grant agreement No 957815.

Conflict of interest

The authors declare no conflict of interest.

Author details

Ivan Todorović* and Ivana Isakov
Faculty of Technical Sciences, University of Novi Sad, Novi Sad, Serbia

*Address all correspondence to: ivan.todorovic@uns.ac.rs

References

[1] V-Model. Available from: https://en.wikipedia.org/wiki/V-Model

[2] Analog Devices. LTspice. Available from: https://www.analog.com/en/design-center/design-tools-and-calculators/ltspice-simulator.html

[3] Sandia National Laboratories. Xyce. Available from: https://xyce.sandia.gov/

[4] POWERSIM. PSIM. Available from: https://powersimtech.com/products/psim/capabilities-applications/

[5] Cadence PCB solutions. PSpice. Available from: https://www.pspice.com/solutions-and-technologies

[6] Typhoon HIL. VHIL. Available from: https://www.typhoon-hil.com/products/virtual-hil-device/

[7] Matlab/Simulink. Available from: https://www.mathworks.com/products/matlab-coder.html".

[8] Plexim. PLECS. Available from: https://www.plexim.com/products/plecs

[9] DIgSILENT. PowerFactory. Available from: https://www.digsilent.de/en/powerfactory.html

[10] PSCAD. PSCAD. Available from: https://www.pscad.com/

[11] EMTP. EMTP. Available from: https://www.emtp.com/products/emtp

[12] MathWorks. C compilers for Matlab. Available from: https://www.mathworks.com/support/requirements/supported-compilers.html

[13] Infineon. Infineon PIL. Available from: https://www.mathworks.com/products/connections/product_detail/infineon-tricore-microcontrollers.html

[14] Infineon. Virtualizer Development Kit. Available from: https://www.infineon.com/cms/en/product/promopages/virtualizer-development-kit/

[15] Opal-RT. ARTEMiS. Available from: https://www.opal-rt.com/solver/

[16] S. N. L. Typhoon HIL. VHIL-Xyce integration. Available from: https://www.typhoon-hil.com/products/xyce-integration/

[17] Typhoon HIL. HIL and Python scripting. Available from: https://www.typhoon-hil.com/documentation/typhoon-hil-software-manual/concepts/script_editor.html

[18] Edrington CS, Steurer M, Langston J, El-Mezyani T, Schoder K. Role of power hardware in the loop in modeling and simulation for experimentation in power and energy systems. Proceedings of the IEEE. 2015;**103**(12):2401-2409. DOI: 10.1109/JPROC.2015.2460676

[19] Feng Z, Peña Alzola R, Seisopoulos P, Guillo Sansano E, Syed MH, Norman P, et al. A scheme to improve the stability and accuracy of power hardware-in-the-loop simulation. In: IECON 2020 the 46th Annual Conference of the IEEE Industrial Electronics Society, Singapore, Singapore, 18/10/20. Piscataway, N.J.: IEEE; 2020. pp. 5027-5032. DOI: 10.1109/IECON43393.2020.9254407

[20] Essakiappan S et al. A multi-site networked hardware-in-the-loop

platform for evaluation of interoperability and distributed intelligence at grid-edge. IEEE Open Access Journal of Power and Energy. 2021;**8**:460-471. DOI: 10.1109/OAJPE.2021.3103496

[21] Available from: http://www.eric-lab.eu/

[22] Available from: https://www.creators4you.energy/

Section 3

Energy Storage Systems and Their Applications in Smart Grids

Chapter 5

Energy Storage Systems and Their Role in Smart Grids

Désiré Rasolomampionona and Mariusz Kłos

Abstract

Energy storage systems play an essential role in today's production, transmission, and distribution networks. In this chapter, the different types of storage, their advantages and disadvantages will be presented. Then the main roles that energy storage systems will play in the context of smart grids will be described. Some information will be given on interactions between energy storage systems and renewables. The emphasis will be on the problems that these storage systems will have to deal with and the possible means that can be used for this purpose. Also the battery management system will be presented as a general concept. The different types of regulation that take place in smart electrical systems (also called smart grids) and the role of energy storage systems will also be discussed. In the end, we will also present one of the biggest weaknesses of storage systems, among others, the degradation of batteries with their use.

Keywords: electric vehicles (EV), energy storage systems (ESS), battery energy storage systems (BESS), wind farms (WF), vehicle-to-grid (V2G), photovoltaic (PV)

1. Introduction

Electrical energy in an alternating current (AC) system cannot be stored electrically. However, there are several methods of its storage by converting AC energy into electromagnetic energy storage systems such as superconducting magnetic energy storage (SMES), electrochemical such as various types of batteries (accumulators), kinetically (flywheels), or even as potential energy (hydropower plants) or as compressed air [compressed air energy storage (CAES)]. The energy storage devices currently available on the market are: battery energy storage systems (BESS), energy capacitor systems (ECS), flywheel energy storage systems (FESS). ESSs in an alternating current (AC) grid cannot store electrical energy directly. **Figure 1** depicts the most important storage technologies for the power grid. Among the devices listed above, the BESS is the most commonly used, but it has drawbacks, such as limited lifetime, current and voltage restrictions, and environmental hazards [1]. As a result of the intensive development of renewable energy sources (RES), the development of electromobility, the need to improve the functioning of the existing power grids, the importance of electricity storage has increased in recent years.

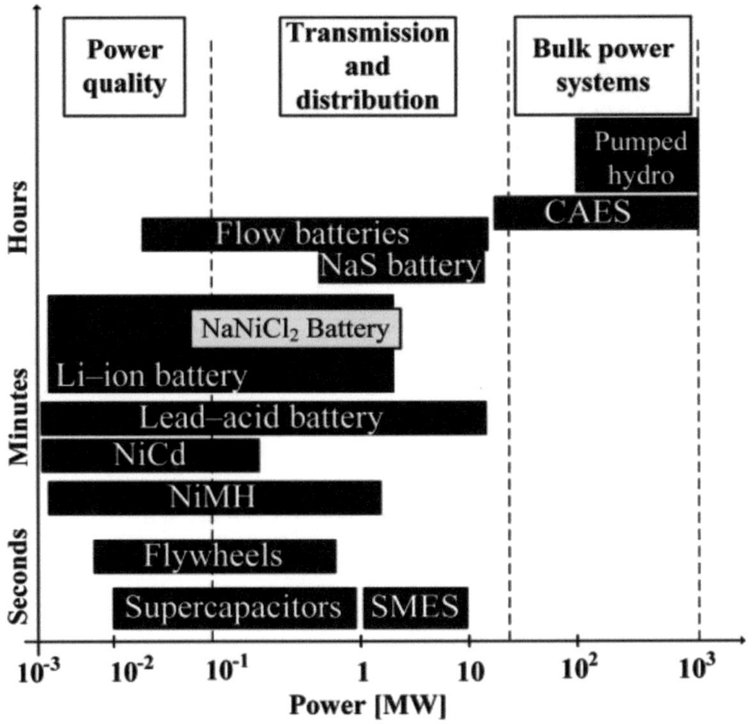

Figure 1.
Storage technologies for the power grid [2].

The superconducting energy storage systems are in the process of moving from their prototype stages to practical applications, which recently also receive special attention for utility applications. The latest technological developments are at such an advanced stage that practically we are now just addressing the performance analyses and the aspect of construction and operating costs. Several articles, among others [3], focus on the performance benefits of adding energy storage to power electronic compensators for utility applications.

Energy storage technologies do not in themselves represent sources of energy. However, they offer significant additional benefits to improve stability, transmission enhancement, power oscillation damping, dynamic voltage stability, tie line control, short-term spinning reserve, load leveling, under-frequency load shedding reduction, circuit break reclosing, subsynchronous resonance damping, power quality improvement, and reliability of supply.

Energy storage systems play a significant role in both distributed power systems and utility power systems. There are many benefits of energy storage systems, including improving the cost-effectivity of the power system and voltage profile. These two features are the most important specifications for storage systems.

Because of the recent development of power electronics, superconductivity, and computer science, the SMES system has received a great attention in the power systems applications. The SMES is notably used in distributed energy storage, spinning reserve, load following, automatic generation control, power quality improvement, reactive power flow control voltage control, and transient stability enhancement [4].

The BESSs have limited lifetime and voltage and current limitations. The FESSs involve other rotating machinery, which is not a preferable, and standby loss is high. Also, the charging method of ECS and its control scheme is not easy.

The fastest-growing power generation technology remains grid-connected solar photovoltaic (PV) power. There was a 70% increase in existing capacity to 13 GW in 2008, while for wind farms, the growth in existing capacity was 29% in 2008 to reach 121 GW, more than double the 48 GW that existed in 2004 [5, 6]. However, like all other renewable energy sources, the main disadvantage of solar and wind energy is their instability. These energy sources depend on natural and meteorological conditions [7]. Great technical challenges related to grid interconnection, power quality, reliability, protection, generation dispatch, and control are to be overcome with higher penetration of intermittent renewable resources [8].

In addition to pumped storage power plants used for years in power systems, other technologies are currently being tested and introduced to enable the storage of electric energy in the form of various energy media.

Electricity storage technologies can be broadly divided into two main categories under the angle of the energy storage form:

- *direct*—it is related to the process of energy storing itself, i.e., how much energy can be stored in a given device. This is a feature of the storage device itself.

- *indirect*—converting energy from electrical to another form (e.g., from electricity to mechanical) determines this. It is also related to the rate at which energy can be transferred to or from the storage device. This conversion mainly depends on the peak power rating of the power conversion unit, but also the response speed of the storage device itself may affect the process.

Type	Energy efficiency (%)	Energy density (Wh/kg)	Power density (W/kg)	Cycle life (cycles)	Self-discharge
Pb-acid	70–80	20–35	25	200–2000	Low
Ni-Cd	60–90	40–60	140–180	500–2000	Low
Ni-MH	50–80	60–80	220	<3000	High
Li-ion	70–85	100–200	360	500–2000	Med
Li-polymer	70	200	250–1000	>1200	Med
NaS	70	120	120	2000	—
VRB	80	25	80–150	>16,000	Negligible
EDLC	95	<50	4000	>50,000	Very high
Pumped hydro	65–80	0.3	—	>20 years	Negligible
CAES	40–50	10–30	—	>20 years	—
Flywheel (steel)	95	5–30	1000	>20,000	Very high
Flywheel (composite)	95	>50	5000	>20,000	Very high

Table 1.
Characteristic parameters of different energy storage technologies.

A detailed overview of various energy storage technologies is presented later in this chapter.

There are several parameters justifying the choice of ESS for an application. One can enumerate the rated power and energy of the application, its response time, its weight, its volume, and its operating temperature. The characteristic parameters of the various energy storage technologies are presented in **Table 1**. These values have been extracted from [9, 10].

2. Different types of energy storage

2.1 Batteries

2.1.1 Lead acid

Lead-acid batteries, commercialized in 1859, are the oldest technology among all batteries that enable the storage of electricity with the use of electrochemical phenomena. Due to its simple structure, the ability to generate high currents, resistance to overcharging, and low price, this technology has become the most common option in DC systems in practically all sectors of the economy. Batteries are used, among others, in automotive starting, lighting and ignition (SLI) and uninterruptible power supplies (UPS), small electric vehicles (e.g., forklifts), or for storing electricity generated in small and medium-sized RES power plants, in the energy and telecommunications sectors.

Due to the maturity of this technology, many new solutions have been developed over the years to optimize the operation of lead-acid batteries, including maintenance-free batteries with liquid electrolyte, with regulated valve (VRLA), with liquid electrolyte absorbed in a separator made of glass mat (AGM), or with gel electrolyte. Despite the design measures that streamline the operation and define new application areas from the point of view of the basic technical parameters, this technology is a technology "leaving" the market. Recent research leads to the conclusion that it is possible to increase power and energy density by replacing lead with lighter materials such as carbon.

2.1.2 Li-ion

Lithium-ion batteries have been used commercially since 1991, primarily to power small electronic devices. In recent years, largely due to the intensive development of electromobility and photovoltaic power plants, the importance of energy storage based on Li-ion cells has increased. Currently, lithium-ion batteries are used both in domestic storage tanks with a capacity of several kilowatt hours and system storage tanks with a capacity of up to several dozen megawatt hours.

Lithium-ion energy storage is characterized by a high voltage of a single battery (usually 3.6 or 3.7 V) and a high energy density. The "power" and "capacity" scaling of the battery tank (as in the case of other battery technologies) consists in combining lithium-ion batteries into series-parallel systems, forming the so-called battery strings.

This type of battery has several advantages; we can list among others the high energy/weight ratios, the absence of memory effect, and the low self-discharge. These batteries find their uses primarily in portable equipment such as laptops, cameras, cell phones, and portable tools. Thanks to its high energy density, Li-ion is also one of the most promising technologies to be used in the power supply of hybrid and rechargeable electric vehicles. However, the start-up costs of the technology remain a fairly significant barrier to its large-scale use.

2.1.3 NiCd/NiMH

The technology of energy storage in nickel-cadmium batteries is known from the beginning of the twentieth century and for many years was the only alternative to lead-acid batteries. Nickel-cadmium batteries are characterized by a short charging time and resistance to ambient temperature fluctuations (from −40°C to +60°C). NiCd batteries were the chemistry of choice for a wide range of high-performance applications between 1970 and 1990. NiCd cells allowed for significant development of portable devices such as radios, camera flashlights, and power tools. The operation of nickel-cadmium batteries is similar to the previously described lithium-ion batteries. In 2006, the Parliament of the European Union approved directives that significantly limit the use of nickel-cadmium batteries. A significant disadvantage of this technology is the occurrence of the so-called memory effect, which causes a decrease in the capacity of the cells during operation. NiCd batteries have also the following disadvantages compared with NiMH batteries: first of all, their life cycle is more expensive. Secondly, in the 1990s, along with the development of lithium-ion and nickel-metal hydride batteries, their role decreased significantly also due to the difficult process of disposal of used batteries, which requires a complex recycling procedure because the batteries contain toxic compounds. This toxicity of Cd, in addition to the lower energy density, and finally the flat discharge curve and negative temperature coefficient could cause thermal runaway during voltage-controlled charging.

For these reasons, nickel metal hydride batteries (NiMH) have gained prominence over NiCd batteries in the recent past. Nickel oxyhydroxide is used by NiMH batteries for the positive electrode and metallic cadmium for the negative electrode. Research on nickel-metal hydride cells began as early as 1967, but initial problems with metal hydride instability led to a greater focus on developing nickel-hydrogen (NiH) technologies. New metal alloys developed in the 1980s allowed for the optimization of NiMH cells and are now widely used as an alternative to disposable alkaline batteries and nickel-cadmium cells, which are characterized by a much lower energy density (about 40% compared with NiMH cells). NiHM batteries have been the chemistry of choice for EV and hybrid EV (HEV) applications due to their relatively high power density, proven safety, good abuse tolerance, and very long life at a partial state of charge.

In the 1990s and 2000s, NiMH was the most popular and mature chemical technology for battery production. Batteries for EVs and hybrid EVs (HEVs) were produced on the basis of NiMH in the 1990s and 2000s. NiMH batteries had relatively high power density, proven safety, good tolerance to abuse, and a very long life at a partial state of charge. The weak point of these batteries was the relatively high self-discharge rate, up to 20% of energy is lost during the first 24 hours after charging, and then 10% during each subsequent month, although the introduction of novel separators has mitigated this problem.

NiMH batteries can also be used in uninterruptible power supply systems (UPS) and in storage tanks cooperating with RES installations. An additional advantage is the possibility of effective recycling and the lack of highly toxic compounds inside the cells, which makes the technology relatively environmentally friendly.

When overcharged, NiMH batteries use excess energy to separate and recombine water. There is then no need to maintain them. However, they should not be charged at such a rate of charge, or cell rupture may occur due to the accumulation of hydrogen. On the other hand, if the battery is overdischarged, the cell may polarize in the opposite direction, which could affect its capacity.

2.1.4 NaS

In sodium-sulfur solid beta alumina, the cathode is made of molten sulfur, the anode is molten sodium, and the electrolyte is a nonporous, solid beta alumina ceramic material (**Figure 2**). As energy is drawn from the energy storage, sodium ions penetrate the solid electrolyte layer toward the cathode, causing the current to flow through the powered circuit. The process is reversed during charging. The battery cells are used to operate in high temperatures (from 300 to 350°C). In 2011–2015, at the University of Kyoto, work was carried out on solutions enabling the operation of sodium sulfur batteries at a much lower temperature of about 100°C, for use in electric vehicles and installations supplying residential buildings.

Due to the possibility of quick entry into operation, high energy density, high efficiency, and long service life, the main area of application of sulfur-sodium batteries is energy storage with very high power and capacity used to optimize the operation of power grids and RES power plants.

2.1.5 FBs

Flow batteries (FBs) production technology is very promising. FBs are produced in such a way that the total energy stored is decoupled from the nominal power. The size of the reactor and the volume of the auxiliary tank are the main elements on which the nominal power and the stored capacity of the battery depend. Thanks to these characteristics, the FB is able to supply large amounts of power and energy required by electric utilities.

One of the most popular FB technologies is the iron-chromium flow batteries (ICBs). This technology was developed in the 1980s by NASA research teams and the Japanese company Mitsui. Thanks to high efficiency of energy exchange (over 80%), easy scaling, and high reliability, ICB cells are a suitable solution for multi-megawatt system energy storage and smaller uninterruptible power supply systems.

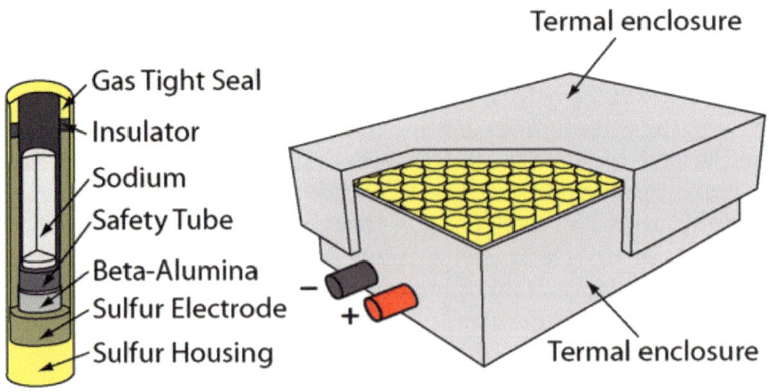

Figure 2.
NaS battery cell and package [11].

ICBs achieve the highest efficiency at relatively high ambient temperatures (in the range from 40°C to 60°C), thanks to which they can be successfully used in regions of hot climatic zones (unlike most electrochemical tanks that require continuous cooling). An additional advantage is the use of common, inexpensive materials—chromium and iron, which are low-toxic, hence safe and environmentally friendly technology.

Energy storage based on ICB cells with a capacity of up to several megawatt hours can be used to secure electricity supplies (e.g., on continental islands or in military bases) and to optimize the operation of RES power plants.

The main disadvantage of ICB cells is the relatively low voltage of a single cell (1.18 V), which results in low energy density and large size of energy reservoirs based on this technology. An additional challenge is posed by the parasitic chemical reactions of chromium with hydrogen, which shorten the life of the cells and cause instability of the liquid electrolyte. It is possible to reduce these phenomena at the expense of lowering the efficiency.

The FB diagram is shown in **Figure 3**. Flow reactants and membrane area define the nominal power, while the total stored energy depends on the capacity of the electrolyte reservoir. It must be remembered that in a conventional battery, the cell itself stores the electrolyte, so there is a strong connection between the power and the nominal energy. A reversible electrochemical reaction takes place in the cell (flow reactor) and produces (or consumes) direct electric current. FB technology is currently used in several large and small-scale demonstration and commercial products.

2.1.6 EDLCs

Supercapacitors, also known as ultracapacitors or electrochemical double-layer capacitors (EDLCs), are characterized by high power density—which translates into short charge and discharge times, high efficiency, and durability. In the case of a supercapacitor, the possibility of quick charging and discharging with high efficiency

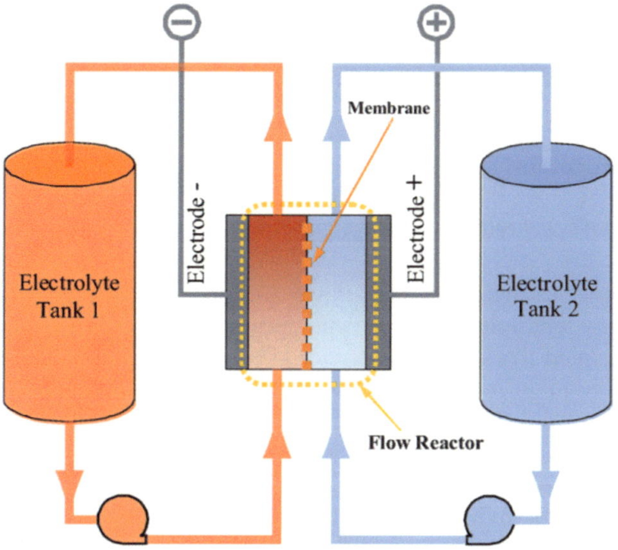

Figure 3.
FB cell.

results from the direct storage of electric energy, because the energy carrier in them is an electric field. Supercapacitors can be used in active filters, improving the quality of electricity, in distribution networks as a tool for energy balancing, and in electric vehicles and trains, popularized, e.g., through Formula 1 racing with KER (kinetic energy recovery systems) systems. There is no faradic process in EDLC, therefore no ionic or electronic transfer results in a chemical reaction. A simple charge separation causes energy to be stored in the electrochemical capacitor.

2.2 Fuel cells

Water and electricity can be produced using FC using electrochemical conversion taking place in special devices that use hydrogen and oxygen. Thanks to the use of FCs, a "hydrogen economy," which is an increasingly popular concept according to which hydrogen is produced by a chemical process, can be ensured. For example, the electrolysis of water, having for objective, among others, the obtaining of hydrogen, which can be used as fuel [12]. Special devices can be used, which combine the function of the FC and the electrolyzer in a single device called regenerative FCs or unitized regenerative FCs. These devices operate as follows: Electricity is produced from hydrogen stored in the form of gaseous fuel, which will later be used for this purpose. FCs are generally optimized to perform only one function, while theoretically they can function as regenerative FCs. By combining the two functions, the size of the system can be reduced for applications requiring both energy storage (hydrogen production) and energy production (electricity production).

2.2.1 Alkaline fuel cells

Alkaline fuel cells (so-called Bacon cells, from the inventor's name—F.T. Bacon) use a liquid alkaline electrolyte (most often potassium hydroxide KOH, which, depending on the type of construction, circulates inside the cell or is contained in an asbestos membrane between the electrodes). An additional advantage of alkaline fuel cells is their resistance to harsh conditions—ambient temperatures below 0°C, high humidity or salt content in the air. AFC cells are currently used primarily as energy sources in uninterruptible power supply systems (UPS), in the own needs of telecommunications and as batteries for electric busses. The disadvantage of AFCs is their low tolerance to carbon monoxide, which reacts undesirably with the electrolyte, making these cells impractical for years.

2.2.2 Phosphoric acid fuel cell

Phosphoric acid fuel cells (PAFCs), developed in the 1960s, were the first commercially produced technology of this type. Since then, PAFC cells have been significantly improved in terms of operational stability, efficiency, and reduction of production costs. In PAFC cells, the electrolyte is gel orthophosphoric acid, placed in a porous layer made of Teflon silicon carbide. The electrodes, on the other hand, are made of porous graphite with an admixture of platinum.

Phosphoric acid fuel cells operate at relatively high temperatures (from 150 to 200°C), which makes them highly resistant to carbon monoxide contamination. Hot water, which is a product of reactions taking place inside the cells, can be used in cogeneration systems for electricity and heat (achieving a high process efficiency of 80%). An additional advantage is their lifetime reaching 40,000 h.

The disadvantage of the PAFC technology is the high corrosivity of the electrolyte, which entails the need to use expensive acid-resistant materials and a relatively low efficiency (30–40%). PAFC cells are used primarily in RES power plants and uninterruptible power systems (e.g., UPS) with installed powers from 50 to 400 kW.

2.2.3 Direct methanol fuel cell

Fuel cells fed directly with methanol are a relatively new solution in the field of electricity storage. The technology was developed by NASA in the 1990s of the last century. DMFC cells use the advantages of methanol as a fuel: high energy density (250–800 Wh/kg), relatively low production costs, and easy transport and storage. This technology is relatively easy to use, because the methanol supplied to the cells can be stored in appropriate tanks located near the bunkers or in replaceable cartridges attached to DMFC cells.

The main area of application of fuel cells directly fed with methanol is loads with relatively low powers, e.g., portable electronic devices or power banks. In recent years, there has been an intensive development of DMFC cells adapted to power small crane vehicles used in large warehouses. Thanks to this, it is possible to shorten the charging time to a few minutes and avoid the costs associated with the installation of battery charging systems for used vehicles.

There are a few other technologies of fuel cell, i.e., Molten Carbonate Fuel Cell, Proton Exchange Membrane Fuel Cell, Solid Oxide Fuel Cell, but their detailed description will be omitted.

2.3 Solar energy and ESS

The annual amount of solar energy received by the earth represents the equivalent of 120,000 TW. Less number of these available solar resources are in a condition to fully replace all nuclear energy and fossil fuels as an energy source [13, 14]. The main obstacles to the further development of solar generation are, among others: the high cost of manufacturing solar cells, dependence on weather conditions, and ultimately, storage and grid connection problem.

Utilities and system operators face some pretty serious challenges due to the integration of significant amounts of solar photovoltaic (PV) generation into the electrical grid. Grid-connected solar photovoltaic units generate and then deliver power to power grids at the distribution level. Installed systems are often designed for one-way power flow from the substation to the customer. The main technical challenges are as follows: transient and steady-state issues due to the widespread adoption of solar generation by customers on the distribution system, voltage variations, sudden weather-induced changes in output, and legacy protection devices designed with power flow in mind [15].

In the case of solar-based electricity generation, weather events such as thunderstorms can have a detrimental effect on solar production—it can range from maximum production to negligible levels in the shortest amount of time. These large-scale weather-related generation fluctuations can be highly correlated within a given geographic area, meaning that the array of solar PV panels on feeder lines downstream of the same substation has the potential to reduce its production considerably in the face of an average meteorological event taking place, for example, on the same day. These disturbances can cause power fluctuations, which can also negatively affect the electrical network in the form of voltage sags if prompt action is not taken to counteract the change in generation. A frequency disturbance can also occur in small electrical systems, resulting from sudden changes in PV generation.

The use of battery energy storage systems (BESS) can provide power quickly in such scenarios to minimize customer interruptions [16] regardless of their location, whether in the center of the substation or distributed along a supply line. Grid-scale BESSs can mitigate the above challenges while improving system reliability and renewable resource economics. This can of course be achieved provided that adequate control schemes are installed.

Regarding the deployment of BESS technologies on the electrical power distribution system, there are two main schools of thought. Centralized storage at the MW level at the distribution station is recommended by a group of scientists. On the other hand, there is a group of people who argue that smaller energy storage systems should be distributed across distribution feeders, networked, and remotely controlled at the substation level.

Each approach has its advantages and disadvantages. Centralized storage has the following advantages in particular—easy access to electrical and Supervisory Control and Data Acquisition (SCADA) equipment of the substation, simplified control schemes, economies of scale on the one hand, and on the other hand because there is already utility-owned land available behind the substation fence. One of the solutions to the problems of deployment of BESS is appropriate sizing and location of the BESS. The ideal sizing and location will depend on the type of site. In the case of large photovoltaic solar installations, preferably a battery system of comparable size connected to the grid is installed in the same substation.

2.3.1 Ramp rate control

One of the main problems with renewable energies is the lack of inertia components. In the case of photovoltaic solar production installations, the inertial components are completely absent. Additionally, the generated power can change very quickly when the sun is obscured by cloud cover. In the case of small electrical systems with high penetration of photovoltaic production, the consequences of this situation could be serious problems of energy supply, since traditional thermal units will have problems compensating for the lack of energy, and hence, maintaining the power balance in the face of rapid changes would be compromised.

As it has been written before, the BESS is used to compensate for the lack of energy in renewable energy installations. In this case where the BESS is coupled to solar power, the BESS must counteract rapid changes in output power to ensure that the installation provides ramp rates deemed acceptable by the system operator. Allowable ramp rates are among the common features of new solar and wind power purchase agreements between utilities and independent power producers. They are usually expressed by the utility in kilowatts per minute (kW/min).

The patent presented in [17] defines the Ramp Rate Control algorithm used in the Xtreme Power - Dynamic Power Resource (XP-DPR) system [18]. This algorithm continuously monitors the actual power output of the solar array and commands the unit to charge or discharge so that the total system power output is within limits set by utility requirements. The operation of an XP-DPR BESS smoothing the volatile power output of a 1 MW solar farm is depicted **Figure 4** [18].

A BESS can be used to discharge when the energy from the solar installation begins to drop in the afternoon. This can be done by charging from the grid at night or from a certain percentage of solar generation during the day. Thanks to this operation,

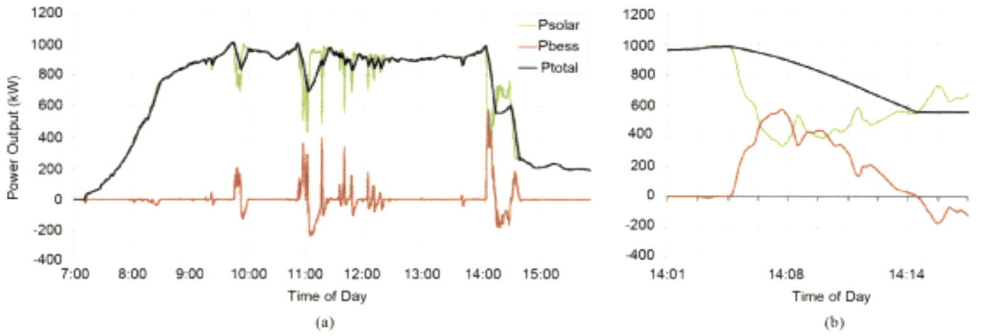

Figure 4.
Ramp rate control to 50 kW/min for a 1 MW photovoltaic installation and a 1.5 MW/1 MWh BESS. (a) Full day. (b) Detail of largest event [18].

the reduction of solar energy at a time when energy is expensive is compensated. This operation is illustrated in **Figure 5** [18].

An excessive part of the electricity produced during the day can be stored in the ESS. On the other hand, during the night, it can be released to complete the energy consumption of a household. Using this method, the total amount of electricity drawn from the power grid can be reduced. Unfortunately, discharging and recharging batteries impact their operational life.

A controller is needed to regulate the charging and discharging process of a battery to protect against overcharging and overdischarging [17]. Despite the use of this controller, some of the energy generated by the PV could still be lost due to the limited energy storage capacity. Apart from this, an inverter must be used to transform the direct current (DC) generated at the PV level into alternating current (AC) to be able to transmit it to the grid. Otherwise, the electrical energy obtained from PV could not be used by household appliances.

In the case of smart grids, it is possible to perform this combination as one of the tasks performed by the main fusion box. Customers choose and register their electricity tariffs in real time, where the price of electricity varies over time [19]. They can also recharge their battery, thereby storing energy in real time from the electricity drawn from the electricity grid, with the aim of reusing it later. The customer can choose the tariff that suits him in order to optimize his bill—for example, by

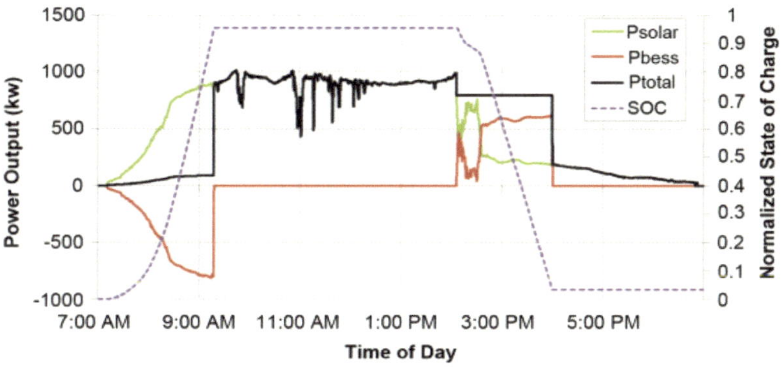

Figure 5.
Full-day output of the solar time-shift application [18].

recharging the battery from the electricity grid when the electricity price is low while discharging it during the period of high electricity prices.

2.3.2 PV and DSM

The scientific literature relevant to energy management in PV-equipped homes mainly focuses on demand-side management (DSM), so how to react to shape the household electrical load during periods of high PV production and to minimize network energy consumption [20].

The local consumption is normally managed in such a way that the ESS battery charge is activated as soon as the PV output power is greater than the electrical load of the house, **Figure 6**. However, this strategy is not able to combat overvoltages that may occur during peak PV production hours (12:00 p.m. to 2:00 p.m.) as the ESS battery is fully charged during the morning hours of sunny days, well before the maximum PV generation period [20].

Figure 6 presents a choice of strategy to be able to move the battery charging period from the "conventional" range to the "proposed" range. This can be achieved by an optimization based on 1-day solar irradiance predictions, described, for example, in [22].

A host of electrical configuration hardware including smart meters, smart sockets, to realize load transfer of different appliances, and main controller to realize load management (i.e., load shift) [23, 24] is used for power management of energy storage systems (ESS) in houses equipped with PV.

2.4 Hydrogen energy storage

The conversion of hydrogen to heat or electricity is fairly easy to accomplish using the popular equation "hydrogen plus air produces electricity and drinking water." Other than that, hydrogen, as the most common chemical element on the planet, is considered an eternal source of energy [14].

In 2004, two institutions, the U.S. National Research Council [25] and the American Physical Society [26], published two comprehensive studies analyzing the technical options concerning the use of hydrogen, including the problem of the cost

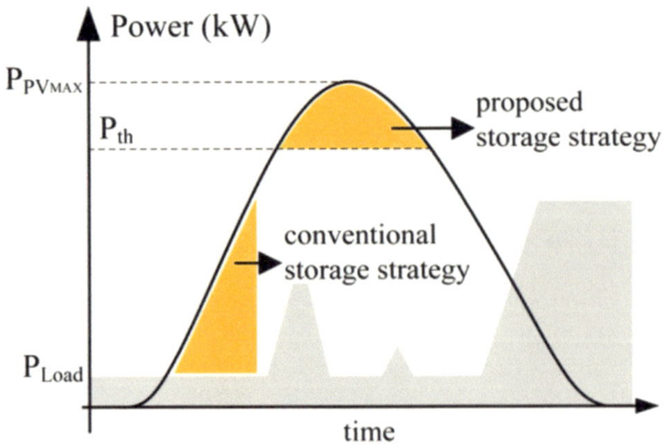

Figure 6.
Conventional storage strategy and proposed strategy compared [21].

of hydrogen obtained from various sources. The only thing missing from these studies is the key question of the overall energy balance of a hydrogen economy.

In fact, there is a whole plethora of processes to be set in motion to obtain hydrogen. We need energy to produce, compress, liquefy, transport, transfer, and store hydrogen. We also lose energy without hope of recovery for its reconversion into electricity with fuel cells [27]. The analysis on the actual energy content in accordance with the law of conservation of energy was analyzed on the basis of the heat of formation or HHV (Higher Heating Value).

2.4.1 How is the hydrogen produced

2.4.1.1 Hydrogen from electrolysis

One of the best-known methods for producing hydrogen is the transformation of water (or rather the dissociation of its molecules) by electrolysis. However, this process is very energy intensive. We envision that in a sustainable energy future, priority will be given to the direct route, i.e., the transformation of renewable electricity into a chemical energy vector. According to [12], the standard water formation potential is 1.48 V, which would correspond to the heat of formation or higher calorific value HHV of hydrogen. The authors [12] also claim that for advanced solid or alkaline polymer electrolyzers, about 0.1 V is lost through biasing, while 0.2 Ω cm^2 is typical for area-specific resistance.

2.4.1.2 Hydrogen generation (PEM electrolyzer) system

The production of hydrogen can be carried out in an efficient manner using the electrolysis of water using polymer electrolyte membrane (PEM) cells. This means of obtaining hydrogen is quite simple to implement. PEM electrolyzers are compact and the current capacity is higher.

The following four auxiliaries are used in the dynamic model of a PEM electrolyzer [28]: the anode, the cathode, the membrane, and the voltage auxiliary (**Figure 7**).

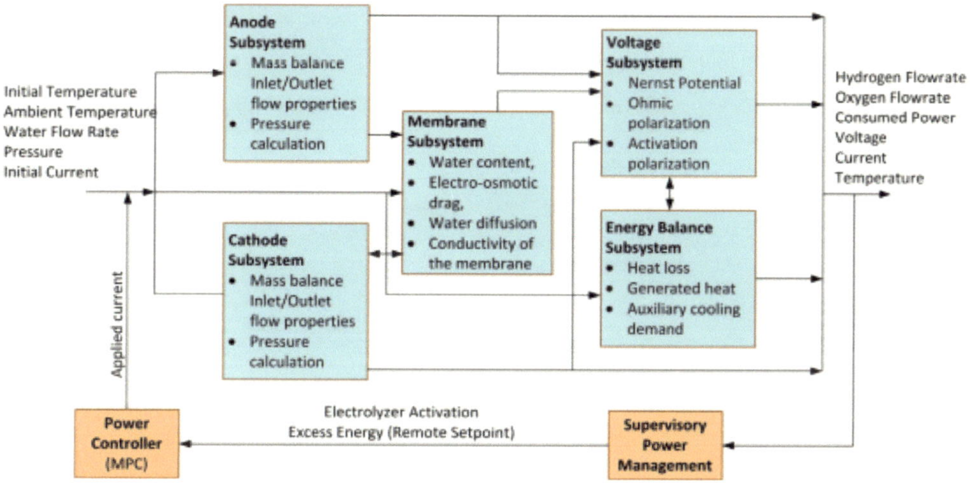

Figure 7.
Electrolyzer modeling block diagram [28].

The flow rates of oxygen and water and their partial pressures are calculated at the auxiliary anode calculated. The calculation of the partial pressures of hydrogen and water as well as their flow rates is carried out by the cathodic system. Water content, electro-osmotic drag, water diffusion, and membrane conductivity are calculated by the membrane auxiliary. The voltage auxiliary calculates the voltage of the electrolyzer by incorporating the Nernst equation, the ohmic bias, and the activation bias.

2.4.1.3 Hydrogen consumption (fuel cell) system

The PEM fuel cell is the inverse equivalent of a PEM electrolyzer. It is modeled similarly to the PEM electrolyzer described in the previous section. A chemical reaction with oxygen is carried out in order to obtain the chemical energy of the hydrogen fuel, which will then be converted into electricity.

Water and heat are the by-products of this reaction. The authors [29] developed the dynamic fuel cell model shown here. This model is made up of four main auxiliaries: the anode, the cathode, the membrane, and the voltage (**Figure 8**).

2.4.1.4 Hydrogen from biomass

It is also possible to produce hydrogen from biomass. However, it seems that this option does not really have a future because, first of all, the process is quite complex: Biomass must be converted into biomethane by aerobic fermentation or gasification before it can produce hydrogen. And secondly, we know, however, that natural-gas-grade biomethane (more than 96% CH4) is already a perfect fuel for transport and stationary applications. Why turn it into hydrogen? There is already a biomethane supply system from waste water digesters in many European countries, the finished

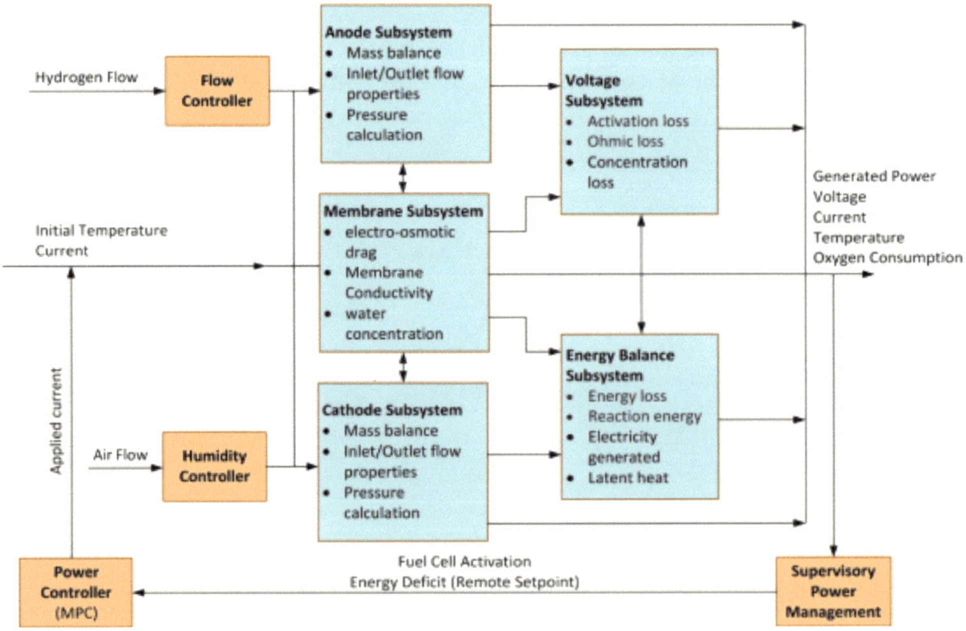

Figure 8.
Full cell modeling block diagram [29].

product of which is already being sold at petrol (fueling) stations to an increasing number of satisfied drivers.

The high energy losses may be tolerated for some niche markets, but it is unlikely that hydrogen will ever become an important energy carrier in a sustainable energy economy built on renewable sources and efficiency.

Moreover, the delivered hydrogen must be converted to a motion for all transport applications. IC engines convert hydrogen within 45% efficiency directly into mechanical motion, while equally efficient fuel cells systems produce DC electricity for traction motors. Further losses may occur in transmissions, etc. All in all, hardly 50% of the hydrogen energy contained in a vehicle tank is converted to motion of a car. The overall efficiency between electricity from renewable sources and wheel motion is only 20–25%.

2.4.2 Hydrogen transformation: fuel-cell-powered vehicles

The most efficient way to use a fuel, in particular hydrogen, in a vehicle is to convert the fuel's energy directly into electricity in a fuel cell. The hybrid design consists of realizing a corresponding illustrated drive train in which the charge of the fuel cell can be leveled using a small battery or an ultracapacitor, much like in a hybrid vehicle with a maintenance electric motor dump. The energy in the battery in question is much smaller than the energy stored in hydrogen. For example, if we store 3 kg of hydrogen, it would be equivalent to three gallons of gasoline or about 100 kWh. This would correspond to more energy than that in the battery of a passenger car.

2.4.2.1 Hydrogen production in micro fuel cell applications

The successful commercialization of miniature fuel cells presents a huge constraint as an alternative to conventional rechargeable batteries for supplying electricity to portable electronic devices such as laptops and mobile phones.

Unfortunately, serious difficulties and significant risks are linked to the storage and handling of hydrogen, whether in the form of compressed gas or liquid, which is used as fuel [30]. Furthermore, compared with storage in the form of liquid hydrocarbons such as methanol, the stored density of hydrogen in compressed or liquid form is significantly lower. This hydrogen can later be reformed to generate the gas when needed. Other methods of hydrogen storage such as in the form of metal hydrides [31, 32] have been discussed extensively in the literature.

There are, however, a number of disadvantages of using hydrides to store hydrogen. These include loss of hydrogen storage capacity after repeated use (limited service life of the alloy), higher weight per unit amount of hydrogen stored (hydrides have the weight of the added metal to the total weight of the storage tank), and the difficulty in extracting all the stored hydrogen due to hysteresis.

2.4.3 Microgrid and hydrogen-based ESS

Most of the projects launched related to wind/solar hydrogen power plant systems [34, 35] and whose results are presented in various scientific articles show the need to introduce greater optimization of the operation of the electrical energy production facility. Although autonomous operation is achieved, several articles report technical problems during operation and serious shortcomings such as electrolyzer breakdowns, high inefficiency in the hydrogen loop, loss of fuel cell performance, breakdowns of pump, etc. It was therefore concluded that the technical problems related to the design and operation of power plants are not yet fully resolved and that an in-depth study is recommended to achieve more reliable operation.

It is also necessary to take into account the fairly complex management of energy production in microgrids (MGs). Energy management in MG is a big challenge to face due to the need to integrate generation, storage systems, and different types of loads, while controlling while the demand is satisfied [36].

In general, two timescales are taken into consideration for MG energy management, as shown in [37]:

- MS energy is analyzed in the long term: this analysis includes generation and load forecasts, maintenance intervals, disconnection of controllable loads and provision of reserve power capacity.

- The authors conducted a short-term energy management of the MG: this analysis takes into account the distribution of energy in real time between the sources and the internal loads.

The process of management of energy production in microgrids is shown in **Figure 9**. The ON-OFF switching thresholds for the electrolyzer and the fuel cell are indicated there. In addition, a protection system against overcharging (high state of charge (SOC)) or undercharging (low SOC) is incorporated into the battery bank.

2.4.4 Control of a grid-connected hybrid system integrating RE, hydrogen, and batteries

There are two uncontrollable but equally essential parameters for the production of RES. These are solar irradiance and wind speed. Therefore, a supporting power source is needed to increase the degree of controllability and operability of the HRES. In almost all solutions for the production and control of renewable energy sources, DC/DC power converters are used to connect them to a central DC bus. In order to coordinate energy use in microgrids, different optimization methods can be used. The use of the supervisory control system based on ANFIS is presented and demonstrated in [38] in order to manage the power of the microgrid.

An example of a grid-connected hybrid system is shown in **Figure 10**. The system is composed of WT and PV panels (renewable and primary energy sources) and a hydrogen subsystem and a battery (SSE).

A three-phase inverter is used to connect the whole system to the grid. Primary renewable sources are generated whenever there is wind or solar radiation. As far as

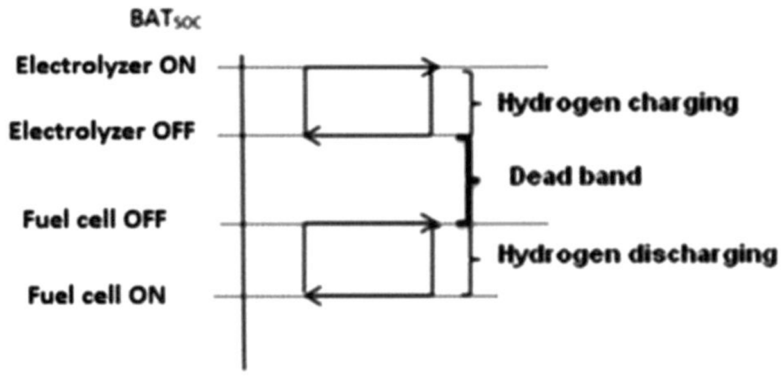

Figure 9.
Scheme of energy management control strategy [33].

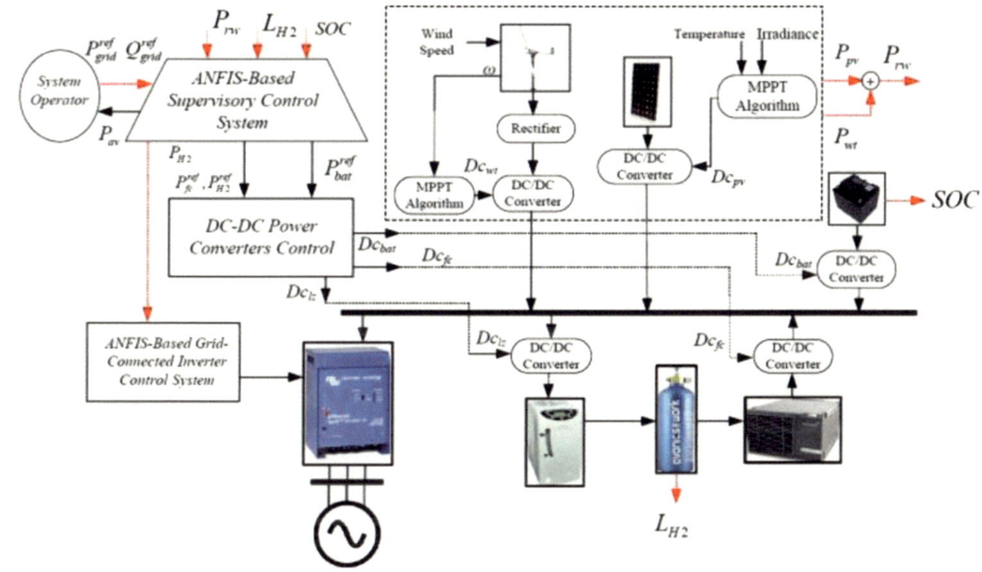

Figure 10.
Grid-connected hybrid system under study [38].

possible, renewable energy is stored in the battery and/or in the form of hydrogen using the electrolyzer. This stored energy will be retrieved at the appropriate time to support renewable generation when needed.

The supervisory control system shown in **Figure 11** is used to determine the power generated by/stored in the hydrogen and the battery. The energy management is carried out taking into account the power requested by the grid, the available power, the level of the hydrogen tank, and the SOC of the battery.

2.5 CAES

In a compressed air energy storage facility (CAES), the surplus energy is used to compress the air for later use. The compressed air is then stored in a cavern as

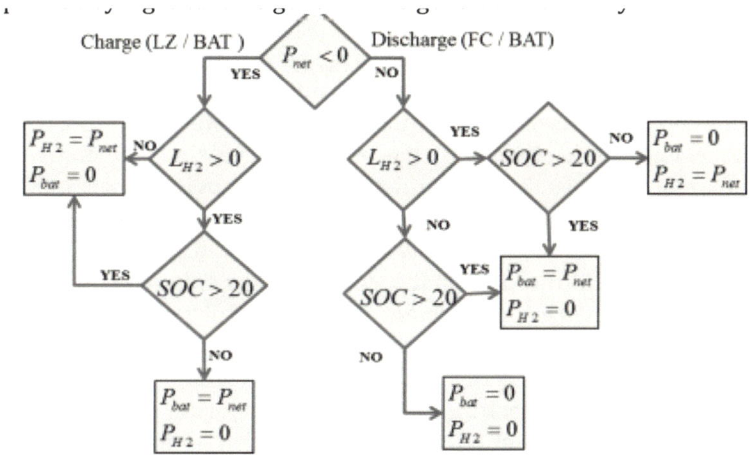

Figure 11.
Scheme of the supervisory control system based on states [38].

potential energy. During the energy demand period, it is expanded back in the turbine. As the energy demand increases, the compressed air is heated and expanded by a gas turbine generator to generate electricity.

CAES replaces the compression ratio of air in the turbine, eliminating the use of fuel gas to compress air. Compression and expansion of air are respectively exothermic and endothermic processes, which in fact makes the design of the system quite complicated. With this in mind, three types of systems are considered to manage heat exchange:

1. Isothermal storage—the air is compressed slowly, allowing the temperature to equalize with the environment [39]. In systems where power density is not critical, such a system is more than enough.

2. Adiabatic systems—in such systems, the heat released is stored during compression and then fed back into the system when air is released. The design of the system is complicated because it requires a heat storage device.

3. Diabatic storage systems—external power sources are used to heat or cool air to maintain a constant system temperature. Most commercially implemented systems are of this type due to the high power density and great flexibility of the system, but at the expense of cost and efficiency.

Many applications are planned for CAES. Among other things, we can cite the use of CAES as a support for the electrical network for load leveling applications [40–42]. In this type of application, energy is stored during periods of low demand and then converted back into electricity when the demand for electricity is high. Natural caverns are used as air reservoirs in commercial systems.

2.6 Flywheel

Flywheel ESS (FESS) is a system for storing energy in a rotating mass, [43].

Flywheel systems are capable of delivering very high peak power. In fact, given recent advances in power electronics and engineering materials, only the power converter is able to limit the input/output peak power. The number of charge-discharge cycles of the FESS is practically infinite. Their power and energy density are very high. Thanks to these characteristics, FESSs are generally used in transmission and power quality applications that require a large number of charge-discharge cycles [44, 45]. This solution is used in particular in synchronous generators to stabilize the output voltage. Lately this technology has become increasingly attractive for a number of other applications such as transmission and improving power quality. FESSs also allow for relatively simple state monitoring, as "state of charge" is a function of easily measurable parameters such as flywheel inertia and speed [46].

The flywheel's maximum rotational speed is the key factor that determines the technology used to build each component. The FESS is classified as low- or high-throughput FESS depending on this speed. The boundary between the two systems is around 10,000 rpm. Not only the material, geometry, and length of the flywheel, but also the type of electric machine and the type of bearing are determined by the rotational speed of the flywheel [40, 47]. High-speed systems are more complex due to technological requirements. However, since the total energy stored in the flywheel depends on the square of the rotational speed, high-speed flywheels provide a higher energy density. Other design considerations such as system performance, security, and reliability are also taken into account [41, 42, 48].

3. Battery management systems

Energy storage systems should intervene in situations where the variation in demand must be taken into consideration. Applications that could benefit from energy storage within the power grid have a wide range of requirements.

There are isolated regions where seasonal energy storage is needed. Megawatt-hours of capacity is stored for months at a time [49]. On the other hand, the stabilization of transport and distribution networks requires that energy can only be stored for a few minutes before being returned to the network or locally. At these precise moments, we are obliged to have energy capacities on the watt-hour scale [50]. Many different forms of energy storage have been developed to operate on all of these time and energy scales. It is also necessary to have an effective management system to maintain safe operation and optimal performance due to the high demands placed on these energy storage systems.

In order to overcome all the different requirements, not only regarding the reaction time of the BESS, a battery management system (BMS) is used to monitor and maintain safe and optimal operation of each battery pack. Additionally, a Supervisory Control System (SSC) must be installed to monitor the entire system.

During their normal operating period, the batteries are permanently in a charge/discharge cycle, therefore in a permanent state of nonequilibrium. Moreover, the situation worsens for the case of storage systems based on intercalation (e.g., Li chemistry), making it difficult to properly monitor battery status and maintain safe operation.

Batteries in a BESS will degrade during cycling, even during normal operation. In extreme load periods, this degradation can even accelerate, especially with the increase in temperature (both ambient and operating). The main role of the basic BMS is to control the batteries only to meet the power demand.

It is possible to reduce the causes of battery degradation and improve system performance by using BMS based on smarter models. There are predictive and adaptive models of BMS, which are particularly useful for large battery packs used in applications such as electric vehicles and grid integration [51–53].

Figure 12 depicts a general BESS-BMS structure for implementing a particular solution used to solve the complex problem of BESS control [54]. The BMS can accurately estimate many internal variables that allow it to gain an in-depth understanding of the battery's state of charge (SOC) and state of health (SOH). This task is carried out using physics-based models.

The tasks for which the BMS is responsible are: operational safety (thermal management, operation between safety current and voltage limits, shutdown on fault detection, etc.), state estimation (determination of the SOC), the estimation of the parameters (determination of the SOH), the remaining time (tr) (according to the load profile applied), and other miscellaneous functions.

For BESSs with Li-ion batteries and other closed-cell systems, the BMS must also perform inter-cell load balancing. For RFBs (redox flow batteries), the BMS must control electrolyte flow based on power demand. Many battery packs with individual BMS will be combined to create a large capacity BESS in large systems. Battery information is transmitted from the BMS to the SSC, which is the interface between the network and the BMS.

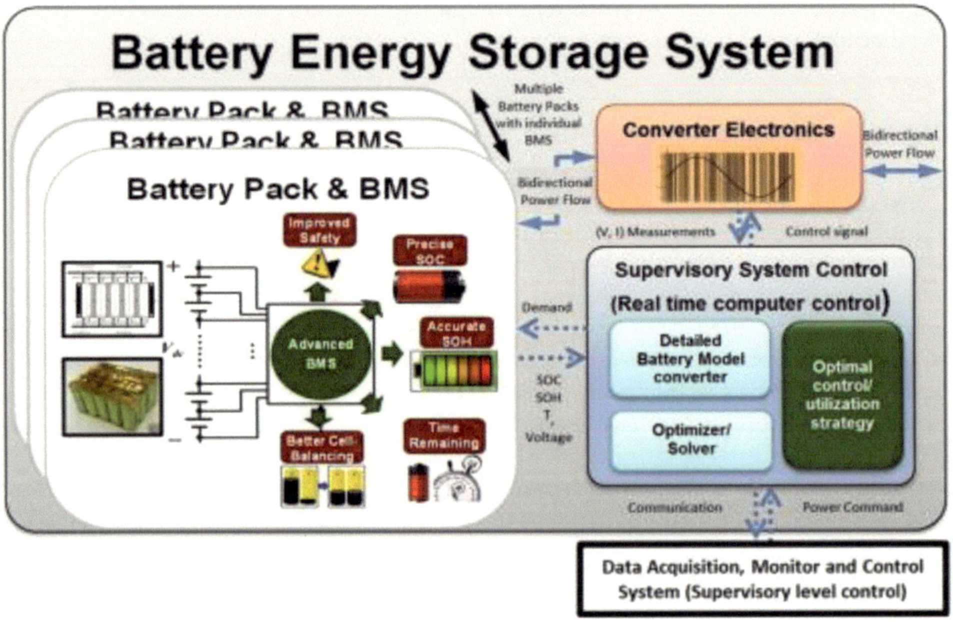

Figure 12.
Schematic for the implementation of a battery pack and BMS into a BESS [54].

The intervention of the BSSs proceeds in the following way—when the grid needs energy from the batteries to supply the load, the SSC chooses the optimal protocol to release the load from a pack (or battery packs) by taking into account both the current state of the batteries and the demand of the network. In order to meet the final power demand, this SSC protocol will call for power to individual packs.

There are times when the required battery power profiles will be more flexible and the BESS may have more control over the charging pattern. For example, the discharge power is severely limited in a peak-shaving application, while the charge power can be chosen according to the needs of the BESS. Here, the best load profiles can be determined by running routines on individual BMSs. The determined load profiles are then transmitted to the SSCs, which then take over the control of the input power of the network.

3.1 BMS architecture

To implement advanced BMS in a grid-scale application requires advanced architecture and a mix of power electronics to connect the battery and BMS within the larger grid. In addition, detailed modeling is extremely useful to predict SOC and SOH as accurately as possible. To manage in real time the nonlinearity, the constraints and the objectives of the model have to be considered. The BMS must be very efficient thanks to the implementation of appropriate algorithms. The implementation of BMS must be done in such a way that an architecture including monitoring and control is realized at several levels [55].

A typical grid storage (GSS) solution consists of a direct current (DC) system, a power conversion system (PCS), a BMS, an SSC, and a grid connection. The DC system is composed of individual cells, which are first assembled into modules, then

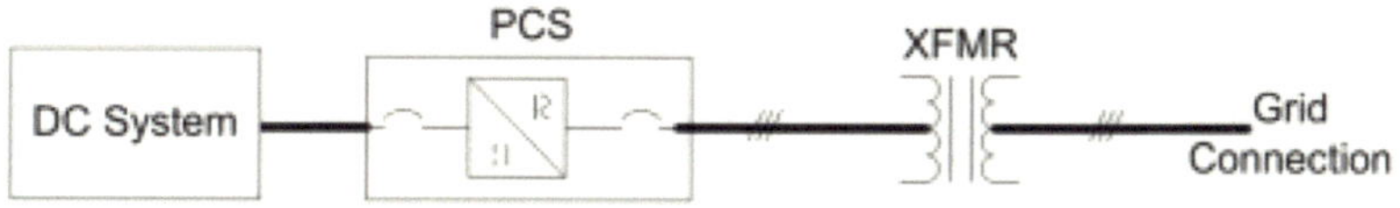

Figure 13.
Simplified illustration of GSS architecture, with a battery-based DC system, a power conversion system (PCS), and a grid connection [54].

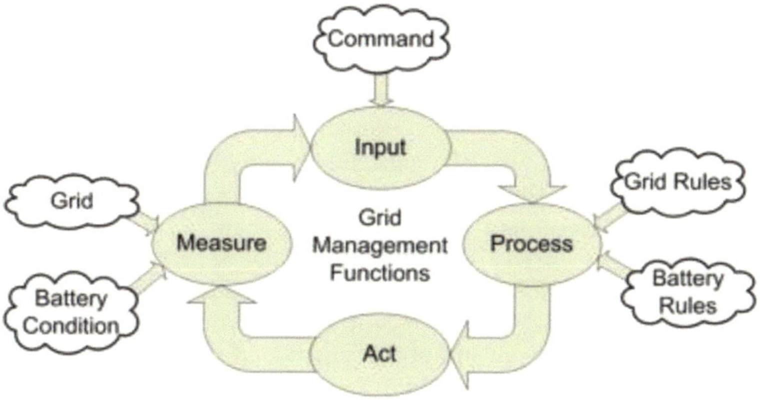

Figure 14.
Conceptual illustration of BMS control cycle [54].

assembled into systems of sufficient capacity to support GSS application requirements. The cells are connected in different electrical configurations in series and in parallel to power a high-voltage bus, which interfaces with the PCS. The PCS is a four-quadrant DC/AC converter connecting the DC system to the grid via a transformer. An illustration of this architecture is shown in **Figure 13**.

Several independent GSSs composed of DC subsystems, PCS, and transformer combinations, called power blocks, can be used in the composition of the system (**Figure 14**). The power supplies can be composed of effectively identical elements, they can also comprise hybrid battery units of different sizes or types. A dedicated BMS manages and controls the operation of the individual power blocks. The SSC on the other hand manages and coordinates the operation of all the power blocks, it also manages the total power of the system and the allocation of this power between the power blocks.

4. Energy storage systems and power grid regulation

As indicated before, high penetration of intermittent renewable resources can introduce technical challenges including grid interconnection, power quality, reliability, protection, generation dispatch, and control. Therefore, the industry will need to confront the challenges associated with higher levels of penetration.

Several articles include a simple diagram for charging and discharging the Battery Energy Storage System (BESS) in order to upgrade the intermittency of renewable energy production. To some extent, this involves storing excess energy when solar/wind power generation exceeds a threshold and offloading it to the grid when load demand is high [56, 57].

4.1 Using ESS for dispatching wind generation

In general, wind energy is considered difficult to control and therefore until now considered non-dispatchable. In conventional grid capacity calculation processes, wind energy is in most cases excluded. One of the suggested ways to overcome this drawback is the use of energy storage systems (see, e.g., [58]). An energy storage system (ESS) can play different roles in the power system—either it can be used to manage energy itself, or it can also be used for energy quality improvement [59].

The combination of energy storage and wind generation improves the availability of wind energy, which can be installed in the grid without worrying about the voltage stability of the system. This allows also to increase the capacity of the existing network infrastructure. Other additional benefits are lower system losses and improved power factor.

For example, the output power of a wind farm can generally be "smoothed" using the ESS, in order to improve the quality of the energy obtained, on the other hand insists on the energy management aspect, which for them is the main objective. In [60], the authors first presented an "*optimization design to determine the most appropriate capacity of the BESS, based on long-term wind speed statistics and maximizing service lifetime/ BESS unit cost*." This is on contrast with what is proposed in [61] where a method for determining the power output schedule of the wind farm, using short-term wind power forecasts, was developed.

Researchers have been taking advantage of the flexible charging/discharging ability of battery energy storage system (BESS) in the design of scheduling schemes for wind farms. In [61], a control strategy for optimal use of the BESS for smoothing out the intermittent power from the wind farm is developed. The simulations the authors have carried out showed that using an actual wind farm data and a realistic BESS model, the desired dispatch set points reasonably close while keeping SOC of BESS within desired limits. The main disadvantage of this method is that it does not allow longer-term power dispatch commitment usually required from generators.

The role assumed by the ESS in wind power trading is another active topic of research (see, e.g., [62]). In some countries, renewable energy plant owners benefit from priority grid supply, where the grid operator has to take control of the energy and pay a fixed return for the energy produced. The article [62] shows how a well-designed ESS can bring additional economic benefits in a project related to energy production. Indeed, if the production of wind energy can be planned in a similar way to the management of energy production from a conventional power plant, the place of wind energy in the power industry will definitely improve significantly, because then one can think about the possibility of dispatching this energy in the power system.

The controllability of power from a wind power generating station by means of BESS is proposed in [63]. This is achieved thanks to two BESSs, one of which is charged using wind power, while the other sends its power into the network. Using statistical wind speed data, the charging characteristics of the BESS are studied and a method to determine the expected charging time of the BESS to reach the stipulated battery state of charge is developed.

A review of design and control of PV and/or wind and/or diesel hybrid systems with energy storage in batteries is presented in [64]. One of the storage technologies to be considered in the future is hybrid hydrogen systems. A particular problem for the installation of this kind of system is the cost of the technologies. Assuming that these costs will decrease over time and investments, the model developed was simulated with prospective costs for the year 2010. The authors of [65] came to the conclusion that the storage system will emerge as the optimal solution, and therefore it is possible to work on a sample of exploitation schemes of a network dominated by renewable energies.

4.2 Use of optimization in designing ESS

As it was written before, one of the main challenges of power generation from wind turbines is the high variability of the wind, which causes the power generation to be intermittent. It is therefore urgent to work on a control system to load/unload the BESS via a converter so that the production of the wind farm can be stabilized and therefore distributed on an hourly basis while taking into account, among others, the SOC and deep discharge limits of the BESS.

This type of problem, whose constraints are more or less difficult to fulfill, can be formulated as an optimal control problem. It is necessary to construct an objective function whose goal is to minimize the deviations between the wind power and the time distribution set points using the BESS. The constraints in question are mainly the constraints on the SOC and the discharge current of the battery, the value margins of which must be fulfilled at all times.

Optimal control, as its name suggests, considers the problem to be solved in such a way as to find a control law for a given system such that a certain criterion of optimality is reached. A control problem includes a cost functional, which is a function of the state and control variables. An optimal control is a set of differential equations describing the trajectories that control variables must follow in order to minimize the cost functional.

Optimal control problems are of different types that can be classified according to (i) the performance index (PI), (ii) the type of time domain (continuous and discrete), (iii) the presence of different types of constraints, and (iv) variables free to be chosen. The optimal control problem can be formulated by considering the following [66]:

1. a mathematical model of the system to be controlled;

2. a specification of the PI;

3. a specification of all boundary conditions on states, and constraints to be satisfied by states and controls.

In [67], an adaptive artificial neural network (ANN)-controlled SMES is presented for enhancing the transient stability of fixed-speed wind farms connected to a multi-machine power system. The control scheme of SMES depends on a sinusoidal pulse width modulation (PWM), voltage source converter (VSC), and DC-DC converter using insulated gate bipolar transistors (IGBTs). An adaptive ANN controller is introduced as the control methodology of DC-DC converter. The effectiveness of the proposed adaptive ANN-controlled SMES is then compared with that of an optimally tuned proportional-integral (PI)-controlled SMES by the response surface methodology and genetic algorithm (RSM-GA) considering both of symmetrical and unsymmetrical faults.

The authors [68] presented a new convex optimization and control method to enhance the value of the lithium-ion-based energy storage system. A novel quadratic objective convex optimization problem, aimed at obtaining an optimal schedule for the BESS, has been elaborated on the basis of technical and economic variables. The objectives of the optimization process, according to authors [68] are: (i) obtaining significantly reduced substation transformer losses, (ii) savings on the cost of energy delivered from the grid, (iii) reducing the life cycle cost of the battery storage system, and finally, (iv) taking into account the variability of the distributed generation resources.

A new method for assessing the role of both WF and ESS is shown in [69]. The main contributions of [69] consist of proposing integrated day-ahead bidding and real-time operation strategies for wind-storage systems (abbreviated WF-ESS) as a price taker. Both the WF and ESS are considered as active actors in the energy market, and their failure to comply with the terms of the contract may result in appropriate penalties. The cooperative strategy is that ESS sets charging or discharging reserve capacities at each time interval up to which the ESS can compensate for potential imbalances from the WF. Coordinating the roles of ESS and WF is to fix the charge or discharge reserve capacity at each time interval to compensate for potential imbalances in WF power production.

4.3 Simulations in wind-storage system studies

Wind systems must be analyzed in a complex way to properly model the phenomena that take place there. Offline electromagnetic transient programs are used, among other things. As a general rule, detailed models take too long when studying slow phenomena, such as the impacts of wind fluctuations on the voltage and frequency of the system, so we tend to build more simplified models, which do not have much impact on the veracity of the results obtained, but on the other hand allow rapid simulations of slow phenomena. However, there may be situations where the loss of detail of specific components in a model can lead to inaccuracy in the simulation, which in turn has significant effects on system design and control testing.

The best way to carry out studies of the ESS system integrated into the wind farm is to do real-time simulations for the following reasons [70]:

1. The wind direction is stochastic. Repetitive simulations of a large number of wind profiles using a large database are sometimes necessary.

2. The nonlinearity present in the characteristics of the equipment as well as the performances of storage of the battery very dependent on the preceding operating conditions makes that the created models are very complex.

3. The WTG and ESS interfaces are modeled considering the detailed Insulated Gate Bipolar Transistor (IGBT) switching bridges. It is also necessary to reproduce as accurately as possible the currents to determine the nominal power and the losses of the SSE when studying the sizing compromises of the total and individual storage elements (the supercapacitor and the battery) under economic, loss and efficiency, and operational constraints.

4. Hardware-in-the-Loop (HIL) tests shall be conducted to verify the prototype ESS controller under various normal and fault conditions.

4.4 ESS and V2G

According to the latest trend resulting from advances in analyzing the impact of human behavior from the perspective of environmental friendliness, electric vehicles can to a certain extent be considered environmentally friendly and can significantly reduce fuel consumption of gasoline. It is predicted that they will dominate the future of the automotive industry.

One of the hottest technologies right now is vehicle-to-grid (V2G) technology that allows electric vehicles to act as distributed energy storages that transfer energy back to the main grid when needed. V2G power flow is achieved by local aggregator through communication between grid and consumers/prosumers. Complete information on power exchange is sent directly from the smart meter to the data centers. The main advantages of V2G include: (a) active support of the network, (b) support of reactive power, (c) control power factor, and (d) support for the integration of RES [71–73]. This effectively achieves load flattening, peak shaving, and frequency regulation throughout a day [74]. Moreover, EVs can even be used to transport energy from remote renewable sources to loads in urgent need of power supply [75]. These services can be obtained by charging the EV during periods of inactivity and add extra EV energy in the electricity grid in peak hours. Apart from supporting the provision of effective power, bidirectional V2G has capability of providing reactive power to ensure the voltage regulation.

On the other hand, there is in inverse operation called G2V, during which the power flows from the generator to the vehicle (G2V) to charge the battery and power flow in opposite to provide peak shaving or concept of "spinning reserve." The power flow can be in any of two modes of operation, namely: (a) unidirectional and (b) bidirectional.

4.5 ESS and ancillary services

V2G electric power capacity can be substantial with attractive ancillary services revenue opportunities. The batteries can act as a source of stored energy to provide a number of grid services. The most promising market for these vehicles is probably that of the ancillary services [76]. Possible services for V2G are: supply of peak power, supply of primary, secondary and tertiary control (for frequency regulation and balancing), load leveling, and voltage regulation. It is unlikely that each vehicle will be contracted separately because the maximum power output of each vehicle is too low. But a fleet manager or aggregator could conclude a contract for a fleet of PHEVs. The advantage of dealing with an aggregator or fleet manager is that a single party represents a more significant amount of power, that is, the accumulated power of the vehicles in the fleet. Moreover, the availability profile of a larger group of vehicles is much smoother. A single vehicle owner could conclude a contract with the aggregator without being concerned about the interface with the electricity markets.

4.5.1 Frequency regulation

The network frequency is one of the most important parameters for evaluating the quality of energy. Frequency regulation is the measure of adjusting the frequency of the system to the nominal value by providing small injections of power (positive or negative) into the network. Many organisms, also called transmission network actors, are responsible for frequency regulation. Examples include Regional Transmission Operators (RTOs) as well as Independent System Operators (ISOs) who simply refer to this service as "regulation."

The theory of frequency control is presented in detail in many books and papers [77–81]. The balance between production and demand between control areas is measured in terms of area control error (ACE). Each control zone generates automatic generation control (AGC) signals based on its ACE values, and the regulation resources respond to the AGC signals to perform the regulation. A complex telemetry system performs this operation in real time and is controlled by the grid operator.

As it was indicated previously, the frequency of the network one of the parameters makes it possible to evaluate the quality of energy. So it is essential to keep the frequency at appropriate levels, i.e., between 49.99 and 50.01 Hz according to the ENTSO-E, the former UCTE [82]. The second equally important parameter is the voltage. Network management consists, among other things, of providing power reserves to maintain frequency and voltage, thus facilitating effective management of imbalances or congestion.

The frequency regulation is carried out at several levels of control: primary, secondary, and tertiary control. Primary Frequency Control (PFC) kicks in during the first few seconds when the system frequency exceeds (or drops under) a pre-established dead band and quickly rebalances the generated and consumed power. In the case of the European power system, the primary reserves regulate the frequency and stabilize the European network to avoid breakdowns. Frequency control is automatically and continuously activated. Primary control can only be activated if primary reserves are available.

Primary regulation is the most demanding in terms of response time and therefore is also the most expensive. This is because PFC has traditionally been provided by thermal generators, which are designed to deliver bulk energy, but not to provide fast-acting reserves. One of the alternative or complementary solutions to the participation of thermal generators in the PFC is the active participation of the loads, which is also considered as a fast and profitable alternative. Nevertheless, a reduction in the load can limit the supply of PFC on the load side because in this case it would prove that the intervention of the load would no longer be very useful.

To complement the generation-side PFC, load-side PFC has been considered as a fast-responding and cost-effective alternative [83–89]. Nonetheless, the provision of load-side PFC is constrained by end-use disutility caused by load curtailment.

In order to balance the network, the secondary reserves are allocated the day before and are adjusted automatically and continuously. The set point is calculated upward and downward on a defined time base (i.e.,15 min) [85]. Most of the research on secondary control of frequency is focused on microgrid stand-alone operation [36, 87]. In this case, we can imagine the intervention of the ESS in the following way: if the frequency is lower than 50 Hz, the batteries could discharge (regulation up), and if the frequency is higher than 50 Hz, the batteries could charge (regulation down). These operations are described in detail in the articles [90, 91]. The frequency can be restored through the use of reserves by leveling out the imbalances between the rated and measured power injections and to restore the frequency.

With regard to tertiary reserves, there are two types: tertiary production reserves and tertiary withdrawal reserves. In both cases, the reserves are used only when major imbalances or major congestions appear. Tertiary reserves are not activated automatically as in the case of primary and secondary reserves, but manually. In practice, the tertiary reserves only intervene very rarely, about a few times a year. Their intervention time is estimated at about 15 min.

So far there is no clear position as to what type or types of ancillary services would be economically profitable for EVs. Scientific opinions differ on the subject. One can quote, for example, the position of the authors [71] who affirm that the secondary and tertiary controls are supposed to be competitive, and the primary control is supposed to be highly competitive. Other authors [92], however, state that the primary control should have the highest value for V2G. Still others say [73] peak power control might be the most economical solution, giving as an example the control system used in Japan. According to an analysis presented in [93], it would seem that the power to be delivered by the tertiary reserves would be too high and the duration too long for the vehicles. So in short, as a compromise, we could say that the possibility of using the primary and secondary controls from the point of view of interaction with the ESS is the most probable.

4.5.2 Voltage regulation

The appearance of an imbalance in the production of electricity can also endanger the voltage, as well as the frequency. In particular, when distributed power generators, such as photovoltaics (PV), increase significantly, a significant increase in voltage may occur due to the reversal of current to the distribution system. In this case, we can highlight the intervention of electric vehicles in the same distribution system in order to absorb the excess electricity, thus minimizing the reverse flow, which will contribute to a balanced electrical condition and a steady voltage.

In a low-voltage grid, the cables are common and contrary to the situation in the transmission networks or the medium-voltage networks, the resistance R is large compared with the reactance X. The adjustment of the active power flow in the grid will influence the magnitude of the voltage. Voltage regulation maintains the voltage within the limits defined by the mandatory standard EN50160 [94]. This voltage control can be integrated into the electric vehicle charger. With regard to the participation of electric vehicles in the regulation of the voltage, it can be carried out in the following way: the load of the vehicles stops when the voltage on the level of the connection to the network becomes too weak. In a later step, the discharge of an active power unit can also be taken into account to increase the grid voltage.

4.5.3 Load leveling and peak power

The electricity load profile generally consists of peak and off-peak loads. Usually electricity suppliers offer different types of tariffs in order to encourage consumers to use the most favorable price ranges during off-peak hours. For load leveling, demand is shifted from peak hours to off-peak hours. Therefore, dispatching is necessary. As in the case of other loads, controllable and aggregate electric vehicles can be discharged during off-peak hours (such as at night and early in the morning), therefore the total load during off-peak hours can be increased, and the gap with the peak hours can be optimized. As the difference between peak and off-peak loads is high, the operation of gensets becomes more difficult, as well as their investment and running costs. Energy stored during off-peak hours is typically released during peak hours to relieve congestion in the grid infrastructure. In this case, peak power delivery and load leveling are the same.

Providing peak power in this way would not be very easy for EVs since the power duration would be relatively long and their storage capacity limited, even in the event that aggregators come into play. On the other hand, from a battery wear point of view, providing peak power is generally not cost-effective as the cost of battery wear would be quite high [95].

Load leveling is more convenient for EVs. The vehicle in this case does not necessarily need to unload during peak hours. Total electricity consumption is simply shifted to off-peak hours of low electricity consumption, which would help minimize power losses and increase grid efficiency. In all these scenarios, the implementation of smart meters or real-time pricing and coordinated pricing is essential, as this would control the incoming and outgoing flows of energy from EVs [96, 97].

4.6 Aggregation of energy storage services through V2G

V2G is defined as the provision of energy and ancillary services, such as regulation or spinning reserves, from an EV to the grid. This can be accomplished by discharging energy through bidirectional power flow or through charge rate modulation with unidirectional power flow [71, 72, 98].

For the vehicle-to-grid concept, three elements are required.

1. First, a power connection to the grid must be available.
2. Second, a control connection is essential for communication with the grid operator.
3. Third, there must be an on-board precision metering for knowing the battery content [99].

The vehicle-to-grid (V2G) estimation methods available in the literature mainly focus on determining the achievable power capacity for a group of EVs [71, 90]. However, these methods are applicable only for determining the V2G capacity and not suitable for real-time V2G capacity estimation and scheduling. Apart from the capacity estimation, other methods have been proposed for aggregating EVs and supplying V2G power to the grid. The aggregation process is also governed by the amount of power and energy that the EVs can supply during any given interval. However, none of the methods available in literature consider dynamic EV scheduling for estimating V2G capacity [90].

In cases where EVs participate in the V2G system, the management, i.e., dispatching, of PHEVs is crucial. Reliable communication must be established between the vehicles and the electrical network, because throughout the duration of the process, data exchange will take place to send the request and carry it out at the level of the EVs. There are three main ways to achieve this communication. First of all, the signal can be sent to each vehicle separately, or via a central controller supervising the EVs, this can, for example, be centralized in a car park. A third possibility is also possible—one can realize the communication using a third-party aggregator, which would be responsible for the separately located vehicles.

Since the energy market system was created, a new player called fleet manager or aggregator has taken place. Its role in the new reality where electric cars are taking an increasingly important place in energy control is to help manage contract systems for a fleet of PHEVs. With vehicle-to-grid (V2G) technology, PEVs parked in a certain area can act as a PEV aggregator when connected to the grid through smart equipment. Such a PEV aggregator can represent a well-defined reactive load and provide additional generation capacity for the provision of ancillary services for power grids [100, 101].

The primary role of the aggregator, a business entity as discussed in [102, 103], is to purchase energy to satisfy transportation needs of its fleet of EVs at the minimum cost. The aggregator is a unit that acts as a mediator between the system operator and individual customers, thanks to which it is possible to coordinate the power exchange between owners of electric vehicles (EV) and the power system. The primary goal of the aggregator is to maximize profits from energy trading and regulatory reserve on wholesale markets. At the same time, the aggregator can also seek to increase their revenue by performing energy arbitrage [102, 104, 105] and/or providing ancillary services [101, 106]. The aggregator itself does not have EV batteries, as they are owned by individual customers who have EVs, therefore they should receive reimbursement of the costs of battery degradation due to their additional use beyond transport needs. The V2G capacities of many electric vehicles are combined by aggregators, then submitted to the appropriate markets [90, 102, 106, 107]. The aggregator in question can be a unit of the public service to which the electric vehicles are connected or a third-party company.

Apart from the fact that an aggregator or a fleet manager is the only interlocutor representing a greater volume of "power," therefore the cumulative power of the vehicles in the fleet is greater, there is also greater flexibility in the supply of power to manage because the greater the number of cars in the fleet, the more fluid the availability of a larger group of vehicles. A single vehicle owner can therefore enter into a contract with the aggregator on more flexible terms without worrying about the interface with the electricity markets.

4.7 Considering the battery degradation

Using the batteries as storage devices for grid purposes reduces their lifetime [105]. Therefore, EV owners must be compensated for the lost utility of their batteries due to degradation when providing services, and this payment will reduce aggregator's revenues. In order for the services from EVs to be economically viable, the revenues must outweigh the cost compensation for the degradation of EV batteries.

Unfortunately, the current state of battery production technology as well as the stage at which scientific research in the field of batteries means that for the moment the use of BESS has a number of drawbacks. The main disadvantages of their use are (i) the large investments, (ii) the associated operating costs due to the degradation of their performance over time (SOH, health status problems). Depending on how BESS is managed, their degradation is increased or mitigated, forcing vehicle owners to replace them after a certain period of time. In this context, the sizing and optimal operation of the BESS are two crucial factors to ensure the extension of their life span necessary to achieve the economic viability of the system.

It has been presented before that aggregators participate in the energy market for a defined commercial purpose, that is, to make the maximum profit. For this purpose, in order to define the annual net benefit (ANP) of the system during the lifetime of a battery, for example, it is necessary to make an estimate of the lifetime of the battery. However, this life span is highly dependent on various operating conditions, and therefore, it is not easy to predict how long a battery will last.

In normal times, the cost of operating the system with a battery system includes two elements: (i) the cost of purchasing electricity (ii) the investment cost of the battery and the inverter. These parameters can be determined in an analytical way presented by example in [108]. However, there are other battery working conditions that are considered "abnormal." We can cite, among other things, the degradation of the batteries during periods when the load demand is exceptionally greater than on other days. Battery storage could increase its profitability by providing fast regulation service under a performance-based regulation mechanism, which better exploits a battery's fast ramping capability. However, battery life might be decreased by frequent charge-discharge cycling, especially when providing fast regulation service. According to [109], it is profitable for battery storage to extend its service life by limiting its operational strategy to some degree. This is also presented analytically among others in [110]. These evaluation methods make it possible to calculate in a close way the loss of capacity of the battery taking into account the discharge rate of the demand during the lifetime of the battery.

Other algorithms were presented for different objectives in order to assure the necessary lifetime extension of ESS so that the economic viability of the system is reached. We can enumerate among others day-ahead forecast errors reduction for wind power, battery energy dispatch, peak shaving, and overvoltage prevention of LV grids, respectively [108, 111].

Authors [112, 113] have presented battery degradation costs associated with additional cycling. The main contributions of the formulation used in [112, 113] are: to simultaneously optimize bidding of V2G, it is necessary to take into consideration: energy, regulation up, regulation down, spinning reserves. They formulated the problem as a linear program, which can be quickly and efficiently solved for large groups of EVs.

Another aspect of the battery degradation is how to compensate the customer's loss. Authors [114] present their point of view concerning this matter. They proposed a bidding strategy for the aggregator to maximize its profits from participating in competitive energy and different regulating reserves markets, while compensating EV owners for degradation. According to [114], an optimal strategy for both energy and reserve markets considering their trade-offs and effect on EV battery degradation has to be taken into account. Also the realistic approach to participating in the voluntary reserve markets with price-quantity offers that are justified is of high priority. And finally assessing the expected profit the EV aggregator can collect by participating in the energy and regulation market is also important.

5. Conclusion

Recent advances in electric energy storage technologies provide an opportunity of using energy storage to address intermittency of renewables. Combining energy storage with renewables improves availability, increases the amount of wind generation that may be installed on the grid without risking the system's voltage stability, increases throughput of existing grid infrastructure, and yields various ancillary benefits.

The solar energy source is the fastest-growing energy source. In small electrical systems, sudden changes in PV generation result in a frequency disturbance; hence, in order to minimize customer interruptions, the use of battery energy storage systems (BESSs) can be of great help.

Hydrogen, as the most common chemical element on the planet, is considered an eternal source of energy. One of the best-known methods for producing hydrogen is the transformation of water. The production of hydrogen can be carried out in an efficient manner using the electrolysis of water using polymer electrolyte membrane (PEM) cells. The inverse equivalent of a PEM electrolyzer is the PEM full cell. It is also possible to produce hydrogen from biomass. However, it seems that this option does not really have a future because the process is quite complex.

All the different requirements regarding among others the reaction time of the BESS can be overcome using a battery management system (BMS), which is aimed at monitoring and maintaining safe and optimal operation of each battery pack is necessary. In addition, a System Supervisory Control (SSC) must be installed to monitor the entire system. It is possible to reduce the causes of battery degradation and improve system performance by using BMS based on smarter models. The BMS can accurately estimate many internal variables that allow it to gain an in-depth understanding of the battery's state of charge (SOC) and state of health (SOH). This task is carried out using physics-based models.

An energy storage system (ESS) can be categorized in terms of the role it plays in a power system: either it is for energy management or for power quality enhancement. Because of the recent development of power electronics, superconductivity, and computer science, the SMES system has received a great attention in the power systems applications. It has been utilized in distributed energy storage, spinning reserve, load following, automatic generation control, power quality improvement, reactive power flow control, voltage control, and transient stability enhancement. As the levels of penetration of renewable energy rise, the technical impact of renewable energy on grid operation led to the application of energy storage for renewables.

Electrical vehicles are among the most popular ESSs, selling energy could be beneficial for EV. Their batteries can act as a source of stored energy to provide a number of grid services. V2G is defined as the provision of energy and ancillary services, such as regulation or spinning reserves, from an EV to the grid. Possible services for V2G are: supply of peak power, supply of primary, secondary, and tertiary control (for frequency regulation and balancing), load leveling, and voltage regulation. In order to make EVs efficiently participate in the regulation process, it is important to know when, statistically, vehicles are available for charging or discharging. The connection to the electric power grid offers opportunities for EVs for charging the vehicle but also for discharging and thus injecting energy into the grid. In order to participate in energy markets, the V2G capabilities of many EVs are combined by aggregators and then bid into the appropriate markets. However, using the batteries as storage devices for grid purposes reduces their lifetime. Therefore, EV owners must be compensated for the lost utility of their batteries due to degradation when providing services, and this payment will reduce aggregator's revenues.

Author note

This chapter could be compiled and published with the financial support of the FUEL GROUP https://www.fueldigital.io/

Author details

Désiré Rasolomampionona* and Mariusz Kłos
Warsaw University of Technology, Warsaw, Poland

*Address all correspondence to: draso@pw.edu.pl

References

[1] Ali MH, Wu B, Dougal RA. An overview of SMES applications in power and energy systems. IEEE Transactions on Sustainable Energy. 2010;**1**(1):38-47

[2] Boicea VA. Energy storage technologies: The past and the present. Proceedings of the IEEE. 2014;**102**(11):1777-1794. DOI: 10.1109/JPROC.2014.2359545

[3] Ribeiro PF, Johnson BK, Crow ML, Arsoy A, Liu Y. Energy storage systems for advanced power applications. Proceedings of the IEEE. 2001;**89**:12

[4] Gustavo MM, Enrique MP. Power flow stabilization and control of microgrid with wind generation by superconducting magnetic energy storage. IEEE Transactions on Power Electronics. 2011;**26**(3):910-922

[5] Renewable Energy Policy Network for the 21st Century (REN21). Renewables Global Status Report 2009 Update [online]. Available from: http://www.ren21.net/pdf/RE_GSR_2009_Update.pdf

[6] Bebic J, Walling R, O'Brien K, Kroposki B. The sun also rises. IEEE Power and Energy Magazine. 2009;**7**(3):45-54

[7] Wang XY, Mahinda Vilathgamuwa D, Choi SS. Determination of battery storage capacity in energy buffer for wind farm. IEEE Transactions on Energy Conversion. 2008;**23**(3):868-878

[8] Abbey C, Joos G. Supercapacitor energy storage for wind energy applications. IEEE Transactions on Industry Applications. 2007;**43**(3):769-776

[9] Sorensen B. Renewable Energy. New York: Academic; 2004

[10] Williams BR, Hennessy T. Energy oasis [vanadium redox battery system in power distribution application]. Power Engineering. 2005;**19**(1):28-31

[11] Energy Storage Association. Available from: https://energystorage.org/why-energy-storage/technologies/sodium-sulfur-nas-batteries/

[12] Bossel U. Does a hydrogen economy make sense? Proceedings of the IEEE. 2006;**94, 10, 1826**:-1837

[13] International Energy Annual 2006. World Energy Reserves, Technical Report by the Energy Information Administration, Official Energy Statistics from the US Government. Report released June December 2008

[14] Rifkin J. The Hydrogen Economy. New York: Tarcher; 2002

[15] Katiraei F, Aguero JR. Solar PV integration challenges. IEEE Power and Energy Magazine. 2011;**9**(3):62-71

[16] Miller N, Manz D, Roedel J, Marken P, Kronbeck E. Utility scale battery energy storage systems. IEEE PES General Meeting; 2010. pp. 1-7. DOI: 10.1109/PES.2010.5589871

[17] "Managing Renewable Power Generation, Xtreme Power," U.S. Patent 20110273129. 2011

[18] Hill CA, Such MC, Chen D, Gonzalez J, Grady WM. Battery energy storage for enabling integration of distributed solar power generation. IEEE Transactions on Smart Grid. 2012;**3**(2):850-857. DOI: 10.1109/TSG.2012.2190113

[19] Mohsenian-Rad A-H, Leon-Garcia A. Energy-Information transmission

tradeoff in green cloud computing. Carbon. 2010. vol. 100. p. 200

[20] Castillo-CagigalaM,Caamaño-MartínbE, Matallanasa E, MasaBoteb D, Gutiérreza A, Monasterio-Huelina F, et al. PV self-consumption optimization with storage and active DSM for the residential sector. Journal of Solar Energy. 2011;**85**(9):2338-2348

[21] Marra F, Yang G, Træholt C, Østergaard J, Larsen E. A decentralized storage strategy for residential feeders with photovoltaics. IEEE Transactions on Smart Grid. 2014;**5**(2):974-981. DOI: 10.1109/TSG.2013.2281175

[22] Lorenz E, Hurka J, Heinemann D, Beyer HG. Irradiance forecasting for the power prediction of grid-connected photovoltaic systems. IEEE Journal of Selected Topics in Applied Earth Observations and Remote Sensing. 2009;**2**(1):2-10

[23] Ganu T, Seetharam DP, Arya V, Kunnath R, Hazra J, Husain SA, et al. nPlug: A smart plug for alleviating peak loads. In: Proceedings of the 3rd International Conference on Future Energy Systems: Where Energy, Computing and Communication Meet. 2012 May:1-10

[24] Shahnia F, Wishart MT, Ghosh A, Ledwich G, Zare F. Smart demand side management of low-voltage distribution networks using multi-objective decision making. IET Generation, Transmission & Distribution. 2012;**6**(10):968-1000

[25] U.S. National Research Council. The hydrogen economy: Opportunities, costs, barriers, and R&D needs [Online]. 2004. Available from: http://www.nap.edu/books/0309091632/html

[26] American Physical Society. The Hydrogen Initiative Panel on Public Affairs [Online]. 2004. Available from: http://www.aps.org/public_affairs/loader.cfm?url=/commonspot/security/getfile.cfm&PageID=49633

[27] Bossel U, Eliasson B, Taylor G. The future of the hydrogen economy: Bright or bleak? In: Proc. Eur. Fuel Cell Forum [Online]. 2003. Available from: http://www.efcf.com/reports/E02_Hydrogen_Economy_Report.pdf

[28] Trifkovic M, Sheikhzadeh M, Nigim K, Daoutidis P. Modeling and control of a renewable hybrid energy system with hydrogen storage. IEEE Transactions on Control Systems Technology. 2014;**22**(1):169-179. DOI: 10.1109/TCST.2013.2248156

[29] Pukrushpan JT, Stefanopoulou AG, Peng H. Control of Fuel Cell Power Systems: Principles, Modeling, Analysis, and Feedback Design. New York, USA: Springer-Verlag; 2004

[30] Pattekar AV, Kothare MV. A microreactor for hydrogen production in micro fuel cell applications. Journal of Microelectromechanical Systems. 2004;**13**(1):7-18. DOI: 10.1109/JMEMS.2004.823224

[31] Ming A, Wu J, Wang Q. The hydrogen storage properties and the mechanism of the hydriding process of some multi-component magnesium-base hydrogen storage alloys. International Journal of Hydrogen Energy. 1995;**20**:141

[32] Tony S, Rangelova V, Nikolay N. Nanocrystallization and hydrogen storage in rapidly solidified Mg-Ni-Re alloys. Journal of Alloys and Compounds. 2002;**334**:219

[33] Valverde L, Rosa F, Bordons C. Design, planning and management of a hydrogen-based microgrid. IEEE Transactions on Industrial Informatics.

2013;**9**(3):1398-1404. DOI: 10.1109/TII.2013.2246576

[34] Ulleberg O, Nakken T, Ete A. The wind/hydrogen demonstration system at Utsira in Norway: Evaluation of system performance using operational data and updated hydrogen energy system modeling tools. International Journal of Hydrogen Energy. 2010;**35**:1841-1852

[35] Yilanci A, Dincer I, Ozturk HK. A review on solar-hydrogen/ fuel cell hybrid energy systems for stationary applications. Progress in Energy and Combustion Science. 2009;**35**(3):231-244

[36] Guerrero JM, Vasquez JC, Matas J, de Vicuna LG, Castilla M. Hierarchical control of droop-controlled AC and DC microgrids—A general approach toward standardization. IEEE Transactions on Industrial Electronics. 2011;**58**(1):158-172. DOI: 10.1109/TIE.2010.2066534

[37] Kanchev H, Lu D, Colas F, Lazarov V, Francois B. Energy management and operational planning of a microgrid with a PV-based active generator for smart grid applications. IEEE Transactions on Industrial Electronics. 2011;**51**(10):4583-4592

[38] García P, García CA, Fernández LM, Llorens F, Jurado F. ANFIS-based control of a grid-connected hybrid system integrating renewable energies, hydrogen and batteries. IEEE Transactions on Industrial Informatics. 2014;**10**(2):1107-1117. DOI: 10.1109/TII.2013.2290069

[39] Lemofouet S, Rufer A. A hybrid energy storage system based on compressed air and supercapacitors with maximum efficiency point tracking (MEPT). IEEE Transactions on Industrial Electronics. 2006;**53**(4):1105-1115

[40] Bolund B, Bernhoff H, Leijon M. Flywheel energy and power storage systems. Renewable and Sustainable Energy Reviews. 2007;**11**(2):235-258

[41] Plater BB, Andrews JA. Advances in Flywheel Energy Storage Systems. PowerPulse.net by Darnell.Com Inc., 2001 March 19. Available from: http://www.darnell.com/NewsServices/PowerPulse/

[42] Furlong ER, Piemontesi M, Prasad P, Sukumar D. Advances in energy storage techniques for critical power systems. In: Proc. BATTCOM. Fort Lauderdale, FL, April 29-May 1 2002. pp. 1-8

[43] Hebner R, Beno J, Walls A. Flywheel batteries come around again. IEEE Spectrum. 2002;**39**(4):46-51

[44] Lawrence RG, Craven KL, Nichols GD. Flywheel UPS. IEEE Industry Applications Magazine. 2003;**9**(3):44-50

[45] Flynn MM, Mcmullen P, Solis O. Saving energy using flywheels. IEEE Industry Applications Magazine. 2008;**14**(6):69-76

[46] EUR 19978 Brochure Energy Storage. A Key Technology for Decentralized Power Power Quality and Clean Transport Office for Official Publications of the European Communities. 2001

[47] Liu H, Jiang J. Flywheel energy storage—An upswing technology for energy sustainability. Energy and Buildings. 2007;**39**(5):599-604

[48] Bernard N, Ahmed HB, Multon B, Kerzreho C, Delamare J, Faure F. Flywheel energy storage systems in hybrid and distributed electricity generation. In: Proc. PCIM. 2003. pp. 121-130

[49] Reuss M, Beck M, Muller JP. Design of a seasonal thermal energy

storage in the ground. Solar Energy. 1997;**59**:247-257

[50] Schoenung SM. Characteristics and Technologies for Long-vs. Short-Term Energy Storage, Sandia National Laboratories. 2001. Available from: https://www.osti.gov/servlets/purl/780306

[51] Verbrugge MW. Adaptive characterization and modeling of electrochemical energy storage devices for hybrid vehicle applications. In: Schlesinger M, editor. Modern Aspects of Electrochemistry 43. New York, NY, USA: Springer-Verlag; 2009

[52] Verbrugge M, Koch B. Generalized recursive algorithm for adaptive multiparameter regression-application to lead acid, nickel metal hydride, lithium-ion batteries. Journal of the Electrochemical Society. 2006;**153**:A187-A201

[53] Verbrugge MW, Conell RS. Electrochemical and thermal characterization of battery modules commensurate with electric vehicle integration. Journal of the Electrochemical Society. 2002;**149**:A45-A53

[54] Lawder MT et al. Battery energy storage system (BESS) and battery management system (BMS) for grid-scale applications. Proceedings of the IEEE. 2014;**102**(6):1014-1030. DOI: 10.1109/JPROC.2014.2317451

[55] Hoff M et al. Multi-configurable, scalable, redundant battery module with multiple fault tolerance. U.S. Patent 7 990 101. 2011

[56] Mehos M, Kabel D, Smithers P. Planting the seed. IEEE Power & Energy Magazine. 2009;**7**(3):55-62

[57] Muljadi E, Butterfield CP, Yinger R, Romanowitz H. Energy storage and reactive power compensator in a large wind farm. In: Proc. 42nd AIAA Aerospace Sciences Meeting and Exhibit. Reno, Nevada; January 5-8, 2004

[58] Yang B, Makarov Y, Desteese J, Viswanathan V, Nyeng P, McManus B, et al. On the use of energy storage technologies for regulation services in electric power systems with significant penetration of wind energy. In: 2008 5th International Conference on the European Electricity Market. 2008. pp. 1-6. DOI: 10.1109/EEM.2008.4579075.

[59] Chen H, Cong TN, Yang W, Tan C, Li Y, Ding Y. Progress in electrical energy storage system: A critical review. Progress in Natural Science. 2009;**19**(3):291-312

[60] Teleke S, Baran ME, Huang A, Bhattacharya S, Anderson L. Control strategies for battery energy storage for wind farm dispatching. IEEE Transactions on Energy Conversion. 2009;**24**(3):725-732

[61] Li Q, Choi SS, Yuan Y, Yao DL. On the determination of battery energy storage capacity and Short-term power dispatch of a wind farm. IEEE Transactions on Sustainable Energy. 2011;**2**(2):148-158. DOI: 10.1109/TSTE.2010.2095434

[62] Voller S, Al-Awaad AR, Verstege JF. Benefits of energy storages for wind power trading. In: 2008 IEEE International Conference on Sustainable Energy Technologies. 2008. pp. 702-706. DOI: 10.1109/ICSET.2008.4747097

[63] Yao DL, Choi SS, Tseng KJ, Lie TT. A statistical approach to the design of a dispatchable wind power-battery energy storage system. IEEE Transactions on Energy Conversion. 2009;**24**(4):916-925

[64] Bernal-Agustion JL, Dufo-Lopez R. Simulation and

optimization of stand-alone hybrid renewable energy systems. Renewable Sustainable Energy Reviews. 2009;**13**(8)

[65] Garcia RS, Weisser D. A wind-diesel system with hydrogen storage: Joint optimization of design and dispatch. Renewable Energy. 2006;**31**(14):2296-2320

[66] Chachuat B. Nonlinear and Dynamic Optimization: From Theory to Practice. Switzerland: Automatic Control Laboratory, EPFL; 2007

[67] Muyeen SM, Hasanien HM, Al-Durra A. Transient stability enhancement of wind farms connected to a multi-machine power system by using an adaptive ANN-controlled SMES. Energy Conversion and Management. 2014;**78**:412-420

[68] Nagarajan A, Ayyanar R. Design and strategy for the deployment of energy storage systems in a distribution feeder with penetration of renewable resources. IEEE Transactions on Sustainable Energy. 2015;**6**:3

[69] Ding H, Pinson P, Hu Z, Song Y. Integrated bidding and operating strategies for wind-storage systems. IEEE Transactions on Sustainable Energy. 2016;7:1

[70] Li W, Joos G, Belanger J. Real-time simulation of a wind turbine generator coupled with a battery supercapacitor energy storage system. IEEE Transactions on Industrial Electronics. 2010;**57**(4):1137-1145. DOI: 10.1109/TIE.2009.2037103

[71] Kempton W, Tomic J. Vehicle-to-grid power fundamentals: Calculating capacity and net revenue. Journal of Power Sources. 2005;**144**(1):268-279

[72] Kempton W, Tomic J. Vehicle to grid implementation: From stabilizing the grid to supporting large-scale renewable energy. Journal of Power Sources. 2005;**144**(1):280-294

[73] Kempton W, Kubo T. Electric-drive vehicles for peak power in Japan. Energy Policy. 2000;**28**:9-18

[74] Zhang YJA, Zhao C, Tang W, Low SH. Profit-maximizing planning and control of battery energy storage systems for primary frequency control. IEEE Transactions on Smart Grid. 2018;**9**(2):712-723

[75] Lam AYS, Leung K-C, Li VOK. An electric-vehicle-based supplementary power delivery system. In: Proc. IEEE Int. Conf. Smart Grid Commun. (SmartGridComm). 2015. pp. 307-312

[76] Clement-Nyns K, Haesen E, Driesen J. The impact of vehicle-to-grid on the distribution grid. Electric Power Systems Research. 2011;**81**(1):185-192. ISSN 0378-7796. DOI: 10.1016/j.epsr.2010.08.007

[77] Elgerd OI, Fosha CE. Optimum megawatt-frequency control of multiarea electric energy systems. IEEE Transactions on Power Apparatus and Systems. 1970;**PAS-89**(4):556-563. DOI: 10.1109/TPAS.1970.292602

[78] Fosha CE, Elgerd OI. The megawatt-frequency control problem: A new approach via optimal control theory. IEEE Transactions on Power Apparatus and Systems. 1970;**PAS-89**(4):563-577. DOI: 10.1109/TPAS.1970.292603

[79] Elgerd OI. Electric Energy Systems Theory, An Introduction. 2nd ed. New York: McGraw Hill Book Company; 1982

[80] Kundur P. Power System Stability and Control. New York: McGraw-Hill; 1994

[81] Machowski J, Bialek JW, Bumby JR. Power System Dynamics: Stability and Control. Chichester, U.K.: Wiley; 2008

[82] UCTE. UCTE Operation Handbook. 2004. Available from: http://www.entsoe.eu/resources/publications/ce/oh/

[83] Zhao C, Topcu U, Li N, Low S. Design and stability of load side primary frequency control in power systems. IEEE Transactions on Automatic Control. 2014;**59**(5):1177-1189

[84] Zhao C, Topcu U, Low SH. Optimal load control via frequency measurement and neighborhood area communication. IEEE Transactions on Power Apparatus and Systems. 2013;**28**(4):3576-3587

[85] Chazarra M, García-González J, Pérez-Díaz J, Arteseros M. Stochastic optimization model for the weekly scheduling of a hydropower system in day-ahead and secondary regulation reserve markets. Electric Power Systems Research. 2016;**130**:67-77

[86] Bidram A, Davoudi A. Hierarchical structure of microgrids control system. IEEE Transactions on Smart Grid. 2012;**3**(4):1963-1976. DOI: 10.1109/TSG.2012.2197425

[87] Shafiee Q, Guerrero JM, Vasquez JC. Distributed secondary control for islanded microgrids—A novel approach. IEEE Transactions on Power Electronics. 2014;**29**(2):1018-1031. DOI: 10.1109/TPEL.2013.2259506

[88] Short JA, Infield DG, Freris LL. Stabilization of grid frequency through dynamic demand control. IEEE Transactions on Power Apparatus and Systems. 2007;**22**(3):1284-1293

[89] Molina-Garcia A, Bouffard F, Kirschen DS. Decentralized demand-side contribution to primary frequency control. IEEE Transactions on Power Apparatus and Systems. 2011;**26**(1):411-419

[90] Sekyung H, Soohee H, Sezaki K. Development of an optimal vehicle-to-grid aggregator for frequency regulation. IEEE Transactions on Smart Grid. 2010;**1**(1):65-72

[91] Rotering N, Ilic M. Optimal charge control of plug-in hybrid electric vehicles in deregulated electricity markets. IEEE Transactions on Power Apparatus and Systems. 2011;**26**(3):1021-1029

[92] Oudalov A, Chartouni D, Ohler C, Linhofer G. Value analysis of battery energy storage applications in power systems. In: 2006 IEEE PES Power Systems Conference and Exposition. 2006. pp. 2206-2211. DOI: 10.1109/PSCE.2006.296284

[93] De Caluwé E. Potentieel van demand side management, piekvermogen en netondersteunende diensten geleverd door plug-in hybride elektrische voertuigen op basis van een beschikbaarheidsanalyse [master's thesis]. K.U.Leuven; 2008

[94] Deilami S, Masoum AS, Moses PS, Masoum MAS. Realtime coordination of plug-in electric vehicle charging in smart grids to minimize power losses and improve voltage profile. IEEE Transactions on Smart Grid. 2011;**2**(3):456-467

[95] Valentine K, Temple WG, Zhang KM. Intelligent electric vehicle charging: Rethinking the valley-fill. Journal of Power Sources. 2011;**196**(24):10717-10726

[96] Ilic MD, Joo JY, Xie L, Prica M, Rotering N. A decision-making framework and simulator for sustainable electric energy systems. IEEE Transactions on Sustainable Energy. 2011;**2**(1):37-49

[97] Hetzer J, Yu DC, Bhattarai K. An economic dispatch model incorporating

wind power. IEEE Transactions on Energy Conversion. 2008;**23**(2): 603-611

[98] Sortomme E, El-Sharkawi MA. Optimal charging strategies for unidirectional vehicle-to-grid. IEEE Transactions on Smart Grid. 2011;**2**(1):131-138

[99] Tomic J, Kempton W. Using fleets of electric-drive vehicles for grid support. Journal of Power Sources. 2007;**168**:459-468

[100] Ma Y, Houghton T, Cruden A. Modeling the benefits of vehicle to- grid technology to a power system. IEEE Transactions on Power Apparatus and Systems. 2012;**27**(2):1012-1020

[101] Sortomme E, El-Sharkawi MA. Optimal combined bidding of vehicle-to-grid ancillary services. IEEE Transactions on Smart Grid. 2012;**3**(1):70-79

[102] Lopes JAP, Soares FJ, Almeida PMR. Integration of electric vehicles in the electric power system. Proceedings of the IEEE. 2011;**99**(1):168-183

[103] Ortega-Vazquez MA, Bouffard F, Silva V. Electric vehicle aggregator/system operator coordination for charging scheduling and services procurement. IEEE Transactions on Power Apparatus and Systems. 2013;**28**(2):1806-1815

[104] Bessa RJ, Matos MA, Soares FJ, Lopes JAP. Optimized bidding of a ev aggregation agent in the electricity market. IEEE Transactions on Smart Grid. 2012;**3**(1):443-452

[105] Ortega-Vazquez MA. Optimal scheduling of electric vehicle charging and vehicle-to-grid services at household level including battery degradation and price uncertainty. IET Generation, Transmission & Distribution. 2014;**8**(6):1007-1016

[106] Quinn C, Zimmerle D, Bradley TH. The effect of communication architecture on the availability, reliability, and economics of plug-in hybrid electric vehicle-to-grid ancillary services. Journal of Power Sources. 2010;**195**(5):1500-1509

[107] Brooks A, Lu E, Reicher D, Spirakis C, Weihl B. Demand dispatch. IEEE Power & Energy Magazine. 2010;**8**(3):20-29

[108] Gitizadeh M, Fakharzadegan H. Battery capacity determination with respect to optimized energy dispatch schedule in grid-connected photovoltaic (PV) systems. Energy. 2014;**65**:665-674

[109] He G, Chen Q, Kang C, Pinson P, Xia Q. Optimal bidding strategy of battery storage in power markets considering performance-based regulation and battery cycle life. IEEE Transactions on Smart Grid. 2016;**7**:5

[110] Hashemi S, Ostergaard J, Yang G. A scenario-based approach for energy storage capacity determination in LV grids with high PV penetration. IEEE Transactions on Smart Grid. 2014;**5**(3):1514-1522

[111] Ru Y, Kleissl J, Martinez S. Storage size determination for grid connected photovoltaic systems. IEEE Transactions on Sustainable Energy. 2013;**4**(1):68-81

[112] Peterson SB, Whitacre JF, Apt J. The economics of using plug-in hybrid electric vehicle battery packs for grid storage. Journal of Power Sources. 2010;**195**(8):2377-2384

[113] Peterson SB, Apt J, Whitacre JF. Lithium-ion battery cell degradation resulting from realistic vehicle and

vehicle-to-grid utilization. Journal of Power Sources. 2010;**195**(8):2385-2392

[114] Sarker MR, Dvorkin Y, Ortega-Vazquez MA. Optimal participation of an electric vehicle aggregator in day-ahead energy and reserve markets. IEEE Transactions on Power Systems. 2016;**31**:5

Chapter 6

Segmented Coil Design Powering the Next Generation of High-efficiency Robust Micro-implants

Yun Tao, Rosti Lemdiasov, Arun Venkatasubramanian and Marshal Wong

Abstract

The next generation of Micro Active Implantable Medical Devices (M-AIMD) are small (< 1 cc), wireless, as well as battery-less. They are located in different parts of the body ranging from brain computer interface electrode arrays (e.g., Blackrock Neurotech Utah Array) to multi-chamber cardiac pacemakers (e.g., Abbott dual chamber Nanostim device). These devices require efficient charging and powering solutions that are very challenging to design. Such solutions require the careful balancing of multiple design parameters such as size, separation distance, orientation, and regulatory limits for emission and tissue safety. In this article, we introduce unique optimisation metrics for designing efficient transmit and receive coils for near-field magnetics-based charging solutions. We elaborate on how the metrics need to be altered depending on the regulatory limits. We discuss the impact of body tissue loading on transmit and receive coil performance using circuit analysis. We introduce a novel "segmented" transmit coil arrangement. We discuss the physics of segmentation, and we build a full wave simulation model, with practical design procedure, which is verified with measurements. Finally, we compare the near fields with and without tissue loading to show that segmented coils offer significant improvement to the performance and robustness of a wireless power transfer system.

Keywords: wireless power, coil, efficiency, delivered power, figure of merit, SAR

1. Introduction

Progress in semiconductor technology has led to the development of substantially miniaturised Micro Active Implantable Medical Devices (also called M-AIMD) that are significantly smaller in size and are implanted in difficult to reach interstitial spaces within the human body, thereby permitting direct interaction with organ systems. This reduction in size facilitates the use of delivery systems (e.g., via catheter or hypodermic needle) that significantly reduce procedure time and burden of care for patients [1–4]. M-AIMDs are either battery-less or have small batteries necessitating the need for efficient charging and powering solutions [5–8]. The most common method is power transferred from an on-body transmitter to an in-body AIMD equipped with a receiver using near-field magnetic induction [9]. This is very challenging as it requires carefully balancing multiple design parameters such as size, depth of implant, orientation of implant (and associated misalignment), and regulatory limits for emission and tissue safety [10]. The need to efficiently deliver power to a small target volume (<1 cc) inside the body requires careful design of the transmit coil system and the receive coil system [11].

There is a lot of work in the literature identifying various parameters that need to be optimised to maximise power delivered to a load (therapy delivering M-AIMD) [12–15]. For example, Fu et al. [16] studied the SIMO (single input multiple output) resonant inductive system and derived an expression for the optimal load and efficiency. Monti et al. [17] concentrated on deriving the solution for the SIMO system that is not necessarily a resonant inductive system. The authors approached the problem of maximising efficiency as a generalised eigenvalue problem. Zargham and Gulak [18] focused their attention on SISO (single input single output) systems. They focused on power transfer through CMOS substrates and lossy biological tissue. Minnaert and Stevens [19] described the three optimisation approaches (efficiency, delivered power and conjugate matching) for SISO systems. Their derivation was based on a generalised 2-port system and was not specific to an inductive resonant system. They suggested that the efficiency of power transmission is a monotonic function of an "extended kQ product, α" which was first introduced by the works of Ohira [20, 21]. Cho et al. [22] studied specific coil designs for wireless power transmission and compared the performance of the designed coils by using a figure of merit defined by Shinohara et al. [23]. In [24], Sharma derived the formulas for efficiency and the figure of merit of a two-coil resonant system. While there are many more relevant articles in the literature, to the best of our knowledge, none of the articles provides metrics to efficiently design the transmitter and receiver coils independently, taking into account the most important regulatory limits for designing these coils for delivering wireless power to medical implants. This is one of the two novel contributions of this article.

This article also focuses on the design of efficient transmit and receive coils where the coil segments are separated by lumped capacitors. These coils are called segmented coils. Segmentation of coils using lumped or distributed capacitors is not new and has been heavily used in Magnetic Resonance Imaging (MRI), for reducing Specific Absorption Rate (SAR) (for transmit coils) [25] and improving coil sensitivity (for receive coils) [26]. Mirbozorgi et al. [27] mentioned that segmentation helps achieve homogeneous power transfer efficiency. Tang et al. [28, 29] stated that the segmentation can significantly reduce the power loss (including the dielectric loss) and required voltage. Stoecklin et al. [30] concluded that capacitive coil segmentation can effectively suppress dielectric losses and non-uniform current distribution. Mark et al. [31] demonstrated that the segmentation results in decrease of the electric field above the transmit coil thereby reducing SAR in the nearby tissue and permitting higher power transfer efficiency. Pokharel et al. [32] use lumped capacitors to segment printed coils and subsequently develop a stacked metamaterial inspired wireless power transfer (WPT) system for efficient and robust power delivery to M-AIMDs. Most of the literature have discussed the positive outcomes of segmentation of coils, but to the best of our knowledge, no one has provided a detailed analytical and numerical (full wave) explanation, as to why segmented coils have lower dielectric losses and significant reduction in SAR when they are near (<1 cm) lossy body tissue. In this article, we address those gaps in knowledge and define our figures of merit (FoMs) to highlight the positive impact of segmenting the transmitter and receiver coils separately. To further validate our novel FoMs, we build and test the transmitter and receiver coils and compare our calculations with measurement results. This is the second novel contribution of this article.

2. Organisation of this work

This article is organised into the following sections:

First, we present a brief overview of the pertinent regulations (exposure and radiation) that limit the performance of WPT systems for medical implants. For a chosen design frequency, we identify the critical parameters that bound the maximum currents that can be carried by a transmit coil and a receive coil. These maximum currents dictate the maximum power that can be delivered by a WPT system.

Second, for a two-coil system, we derive, using circuit analysis, the optimal load resistance needed to maximise (a) delivered power and (b) efficiency. For both cases, we find the receive coil current, delivered power and efficiency.

Third, we derive an optimisation metric we term, *system figure of merit*, for a two-coil WPT system and show that the popular system link efficiency used in the literature, is a monotonic function of the system figure of merit. We split the system figure of merit into two parts: transmit figure of merit (characterising the transmit coil) and receive figure of merit (characterising the receive coil). We demonstrate that an increase of any of these two figures of merit results in an increase of the overall system link efficiency.

Fourth, we identify two mechanisms that cause proximity of lossy dielectric tissue to impact the impedance of a transmit coil. We identify the first mechanism to be associated with the interaction of the coil current with the tissue and the second mechanism to be associated with the interaction of the charges accumulated in the coil with the tissue.

Fifth, we study the effects of introducing lumped capacitors in series with coil wiring to break the coil turns into segments. We study the effect of segmentation on the resistance and reactance of coils. We investigate the impact that segmentation capacitors have on the transmit and receive figure of merit of the coil.

Finally, we validate our circuit models and associated transmit and receive figures of merit with measurements and full wave simulations in HFSS. We perform measurements and full wave simulations of the electrical properties of transmit coil design to demonstrate that the introduction of the segmentation capacitors improves the figure of merit of the transmit coil when it is both unloaded (in air) and loaded with lossy tissue.

3. Review of regulations

Figure 1 diagrammatically illustrates that the regulations associated with wireless power transmission to an implant from an external transmitter can be divided into two groups: radiation (EMC) and exposure. We note that in this article we have examined the regulatory limits for USA and Europe only.

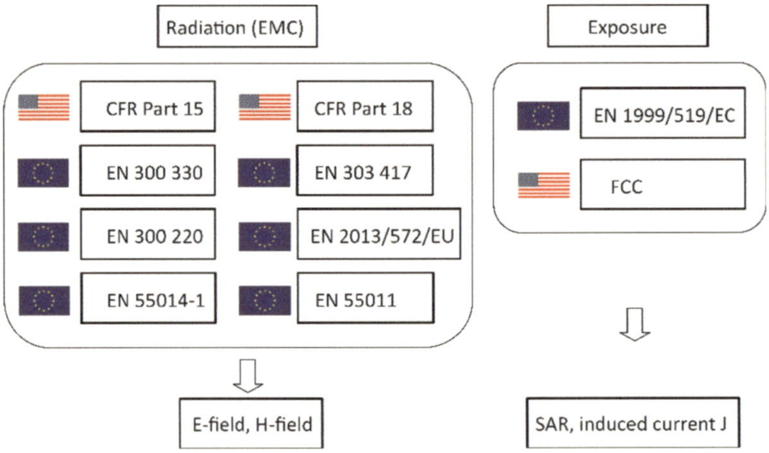

Figure 1.
Diagrammatic representation of the pertinent FCC and EU regulations.

3.1 Exposure

Ensuring the safety of the human body during exposure to electromagnetic waves is an indisputable fact. SAR is a measuring factor for electromagnetic wave absorption. SAR is calculated as

$$SAR = \frac{\sigma_{tissue}}{\rho_{tissue}} |E|^2 \tag{1}$$

where σ_{tissue} is the conductivity of the tissue in S/m, ρ_{tissue} is its mass density in kg/m^3, and |E| is the RMS magnitude of the induced electric field in the tissue due to exposure to these EM waves. FCC [33] limits the peak average SAR to 1.6 W/kg, averaged over 1 gram of tissue. EN 1999/519/EC [34] limits SAR, as well as the volumetric current in the tissue. The actual limit values of SAR depend on the body part exposed to the RF energy. The limit on the induced current depends on the frequency.

3.2 Radiation

The FCC rules and regulations are presented in Title 47 of the Code of Federal Regulations (CFR). Part 15 [35] covers the radio frequency devices. Part 18 [36] covers the Industrial, Scientific and Medical Equipment (ISM). Part 15 and Part 18 limit the radiated electric field at 3 m or 30 m depending on frequency.

EN 300330 [37] covers Short Range Devices (SRD) in the frequency range 9 kHz to 25 MHz and inductive loop systems in the frequency range 9 kHz to 30 MHz. It is a harmonised standard covering the essential requirements of article 3.2 of Directive 2014/53/EU. The standard limits the magnetic field at 10 m from the device. The most generous H-field limits are in three frequency bands containing 6.78 MHz, 13.56 MHz, 27.12 MHz.

EN 303417 [38] covers the wireless power transmission systems, using technologies other than radio frequency beam in the 19–21 kHz, 59–61 kHz, 79–90 kHz, 100–300 kHz, 6765–6795 kHz ranges. It is harmonised standard covering the essential requirements of article 3.2 of Directive 2014/53/EU.

EN 300220–2 [39] covers SRDs operating in the frequency range 25–1000 MHz. for non-specific radio equipment. The most generous H-field limits are in two frequency bands containing 27.12 MHz, 40.68 MHz.

EN 2013/572/EU [40] covers SRDs too. The emphasised frequency bands having higher limits are centred at 6.78 MHz, 13.56 MHz, 27.12 MHz, 40.68 MHz.

EN 55014–1 ("CISPR 14") [41] covers household appliances, electric tools and similar apparatus. This regulation is very restrictive (3 dBμA/m at 3 m in 4–30 MHz range) when applied to the inductive loops and WPT devices.

EN 55011 ("CISPR 11") [42] covers ISM equipment. The devices are sorted into two groups (Non-ISM and ISM equipment) and two classes (non-residential environment and residential environment).

4. Derivation for delivered power and efficiency

Most commonly, WPT circuits use electromagnetic coupling between coils. These WPT circuits use capacitors to reduce reactive power. **Figure 2** is a commonly chosen series–series capacitor representation which has been widely used because the capacitances can be chosen independent of the load and coupling conditions.

At resonance, $\omega^2 = \frac{1}{L_t C_t} = \frac{1}{L_r C_r}$, the equation that links the currents in the transmit and in the receive coil is:

$$j\omega M I_t + (R_r + R_L) I_r = 0 \tag{2}$$

where I_t is the current in the transmit coil, I_r is the current in the receive coil, R_r is the resistance of the receive coil, R_L is the resistance of the load, and M is the mutual inductance. In our derivations, we assume that the voltages and currents are strictly sinusoidal. Therefore, we replace the time derivatives $\partial/\partial t$ by $j\omega$ (multipliers).

From Eq. (2) we can see that there is a 90-degree phase shift between the transmit and receive currents. The $I_{t,limit}$ and $I_{r,limit}$ are the maximum allowed currents in the transmit coil and in the receive coil, correspondingly.

If we ignore the phase shift and redefine jI_r as new I_r, then we get the following expression for the load resistance:

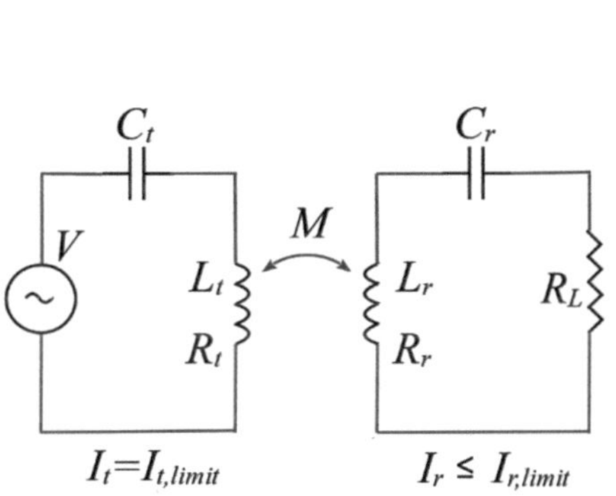

Figure 2.
Schematic of the WPT system.

$$R_L = \frac{\omega M I_t}{I_r} - R_r \tag{3}$$

Additionally, the delivered power is:

$$P_L = \frac{1}{2} R_L I_r^2 = \frac{1}{2}\left(\omega M I_t I_r - R_r I_r^2\right) \tag{4}$$

The power loss in the receive coil is:

$$P_r = \frac{1}{2} R_r I_r^2 \tag{5}$$

The power loss in the transmit coil is:

$$P_t = \frac{1}{2} R_t I_t^2 \tag{6}$$

The efficiency is:

$$\eta = \frac{P_L}{P_L + P_r + P_t} = \frac{1 - \frac{R_r}{\omega M}\frac{I_r}{I_t}}{1 + \frac{R_t}{\omega M}\frac{I_t}{I_r}} \tag{7}$$

We assume that the current in the transmit coil is fixed at its maximum value of $I_t = I_{t,limit}$.We can proceed with two ways: (a) to maximise the delivered power and (b) to maximise the efficiency.

(a) Maximising the delivered power.

We differentiate the delivered power with respect to I_r, equate it to zero and obtain the optimal current in the receive coil:

$$I_{r,opt} = min\left(\frac{\omega M}{2R_r} I_t, I_{r,limit}\right) \tag{8}$$

where the receive current is limited by $I_{r,limit}$.

(b) Maximising the efficiency.

We differentiate the efficiency, equate it to zero and obtain the optimal current in the receive coil:

$$I_{r,opt} = min\left(I_t \cdot \frac{R_t}{\omega M}\left(\sqrt{1+Q_M^2} - 1\right), I_{r,limit}\right) \tag{9}$$

where

$$Q_M = \frac{\omega M}{\sqrt{R_t R_r}}, \tag{10}$$

which we call a mutual quality factor.

We use the optimal receive current to obtain the expressions for the optimal delivered power P_L by using Eq. (4) and efficiency η by using Eq. (7).

If the current limits are high (infinite), then both cases can be elaborated further:

a. Maximising the delivered power, with high current limits.

Optimal current is:

$$I_{r,opt} = \frac{\omega M}{2R_r} I_t \tag{11}$$

Optimal load resistance is:

$$R_{L,opt} = R_r \tag{12}$$

Delivered power is:

$$P_L = \frac{1}{8} \frac{(\omega M)^2}{R_r} I_t^2 \tag{13}$$

Efficiency is:

$$\eta = \frac{1}{2 + \frac{4R_t R_r}{(\omega M)^2}} = \frac{1}{2 + \frac{4}{Q_M^2}} \tag{14}$$

b. Maximising the efficiency, with high current limits.

Optimal current is:

$$I_{r,opt} = \frac{R_t}{\omega M} \left(-1 + \sqrt{1 + Q_M^2} \right) I_t \tag{15}$$

Optimal load resistance is:

$$R_{L,opt} = R_r \sqrt{1 + Q_M^2} \tag{16}$$

Delivered power is:

$$P_L = \frac{1}{2} \frac{R_t R_r}{\omega M} I_t^2 \left(1 + Q_M^2 - \sqrt{1 + Q_M^2} \right) \tag{17}$$

Efficiency is:

$$\eta = \frac{\sqrt{1 + Q_M^2} - 1}{\sqrt{1 + Q_M^2} + 1} \tag{18}$$

From the above formulas we see that Q_M serves as the system figure of merit. Increase of the Q_M leads to the increase of the efficiency of the transfer in both cases.

5. Figures of merit for transmit and receive coils

5.1 From efficiency perspective

For some WPT systems the magnetic field of the transmit coil is not changing significantly in the region of space that contains the receive coil. This happens if the receive coil is much smaller than the transmit coil and/or it is located far enough from the transmit coil. The induced voltage of the receive coil is $j\omega MI_t$ (where M is the mutual inductance). However, according to Faraday's law of induction, the induced voltage is $j\omega B_t A_r$, where B_t is the magnetic field of the transmit coil, A_r is the area of the receive coil.

The expression for the mutual quality factor can be expressed as follows:

$$Q_M = \frac{\omega M}{\sqrt{R_t R_r}} = \omega \frac{B_t/I_t \cdot A_r}{\sqrt{R_t R_r}} = \omega \frac{B_t/I_t}{\sqrt{R_t}} \cdot \frac{A_r}{\sqrt{R_r}} \tag{19}$$

We observe that the values of the transmit and receive coil can be separated. We can define the figures of merit for the transmit coil F_t and the receive coil F_r

$$F_t \equiv \frac{B_t/I_t}{\sqrt{R_t}} = \frac{B_t}{\sqrt{2P_{in}}} \tag{20}$$

$$F_r \equiv \frac{A_r}{\sqrt{R_r}} \tag{21}$$

where $P_{in} = \frac{1}{2}R_t I_t^2$ is an input (transmitted) power into the transmit coil.

The expression for the receive figure of merit F_r can be represented differently in the following way. From the expression for the delivered power (13):

$$P_L = \frac{1}{8}\frac{(\omega B_t/I_t \cdot A_r)^2}{R_r} I_t^2 = \frac{1}{8}\omega^2 B_t^2 \cdot \frac{A_r^2}{R_r} \tag{22}$$

we get:

$$F_r \equiv \frac{A_r}{\sqrt{R_r}} = \frac{\sqrt{8P_L}}{\omega B_t} \tag{23}$$

From Eq. (19) we get the expression for the mutual inductance as:

$$Q_M = \omega \cdot F_t F_r \tag{24}$$

Increase in any of these two figures of merit (F_t and F_r) leads to an increase in the efficiency. The mutual quality factor Q_M can be seen as a figure of merit for the transmit-receive coil system. The system with higher Q_M is more efficient.

The expressions for the transmit figures of merit defined as $F_t = \frac{B_t}{\sqrt{2P_{in}}}$ and $F_r = \frac{\sqrt{8P_L}}{\omega B_t}$ are more general than those defined using the transmit resistance R_t, receive resistance R_r and receive coil area A_r. These definitions apply not only to coils, but also to any "structure" that can perform the following tasks: (a) generate magnetic field (if transmit structure), (b) harvest RF energy (if receive structure). The transmit and receive figures of merit are defined as follows:

- Transmit figure of merit $F_t = \frac{B_t}{\sqrt{2P_{in}}}$ is a measure of ability of the transmit structure (or coil) to generate the RF magnetic field given the consumed input power.

- Receive figure of merit $F_r = \frac{\sqrt{8P_L}}{\omega B_t}$ is a measure of ability of the receive structure (or coil) to harvest the RF power from the incident magnetic field.

It is worth mentioning that the F_t and F_r figures of merit are not the properties solely of the transmit coil and receive coil correspondingly. The coil resistances (and consequently the figures of merit) are affected by the nearby tissue. The coil-tissue separation distance clearly affects these figures of merit. These figures of merit also depend on frequency.

5.2 From delivered power perspective

The expression for the delivered power Eq. (22) can be modified as follows:

$$P_L = \frac{1}{8}\omega^2 \cdot B_t^2 \cdot F_r^2 \tag{25}$$

The delivered power is proportional to the square of the receive figure of merit F_r. It is also proportional to the square of the magnetic field B_t of the transmit coil. This seems to be an intuitive result: the higher the magnetic field is, the more power we can harvest from it.

5.2.1 Considering SAR limit

The magnetic field that we are able to generate at the location of the receive coil cannot be arbitrarily high: the current in the coil is limited by exposure and radiation limits. SAR limit is one of these limits. One can define a SAR figure of merit as a ratio of the magnetic field of the transmit coil to the square root of SAR:

$$F_{t,SAR} \equiv \frac{B_t}{\sqrt{SAR}} \tag{26}$$

By defining the SAR figure of merit using Eq. (26) the maximum achievable magnetic field would be calculated as $F_{t,SAR} \cdot \sqrt{SAR}$. The FCC limit of SAR is 1.6 W/kg.

It is worth saying that the $F_{t,SAR}$ figure of merit is not a property solely of the transmit coil. It is a property of the combination of the transmit coil and the nearby tissue. The coil-tissue separation clearly affects the $F_{t,SAR}$. This figure of merit is also a function of frequency.

This figure of merit can also be used to compare the competing designs of the transmit coils. The transmit coil with higher $F_{t,SAR}$ can deliver more power to the receive coil.

5.2.2 Considering other limits

Apart from SAR, there are other regulations that limit the transmit coil current and the transmit coil magnetic field. For each one of them one can establish the corresponding figure of merit in the following way:

a. Volumetric current J, according to EN 1999/519/EC, if below 10 MHz. The corresponding figure of merit would be:

$$F_{t,J} \equiv \frac{B_t}{J} \tag{27}$$

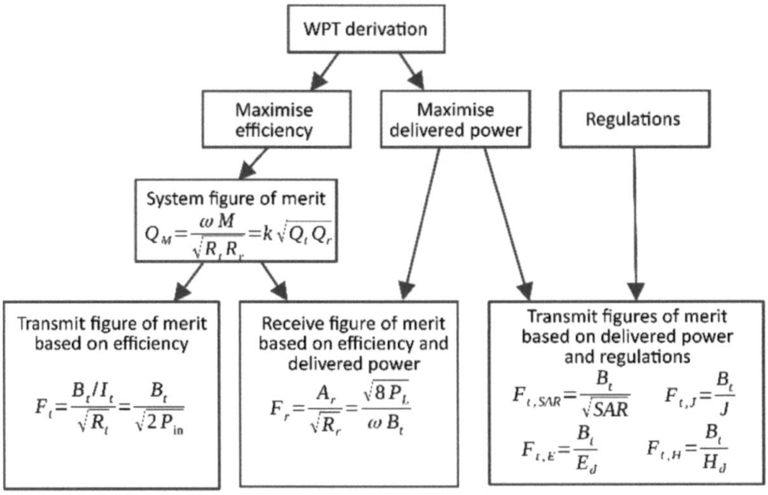

Figure 3.
Figures of merit diagram.

b. Electric field at the certain distance from the coil (d = 3 m, 30 m, 300 m). The corresponding figure of merit would be:

$$F_{t,E} \equiv \frac{B_t}{E_d} \tag{28}$$

c. Magnetic field at the certain distance from the coil (d = 10 m). The corresponding figure of merit would be:

$$F_{t,H} \equiv \frac{B_t}{H_d} \tag{29}$$

Figure 3 provides a visual representation of the development of figures of merit from the WPT formulas and the regulations.

6. Impact of tissue loading on the transmit and receive coils

The electric field of the transmit coil can be separated into two parts: the "current" electric field and the "charge" electric field:

$$\mathbf{E} = \mathbf{E}_{current} + \mathbf{E}_{charge} = -j\omega\mathbf{A} - \nabla\Phi \tag{30}$$

where $\mathbf{A}$ is the magnetic vector potential and Φ is the electric scalar potential. **Figure 4** shows the two components of the electric field when a WPT coil is close to lossy dielectric tissue (e.g. muscle).

The $\mathbf{E}_{charge}$ mostly exists between the terminals of the coil. The $\mathbf{E}_{current}$ electric field exists as concentric circles above the coil.

6.1 $\mathbf{E}_{current}$ electric field

$\mathbf{E}_{current}$ electric field infiltrates the tissue and excites current in it. The current in the tissue flows in self-terminating lines as shown in **Figure 5**. This leads to ohmic losses

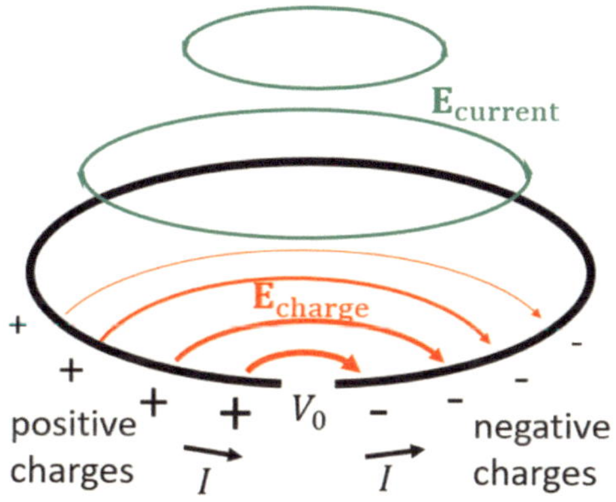

Figure 4.
Electric field of the coil.

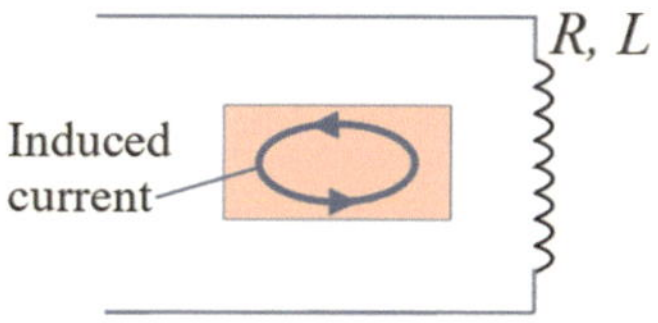

Figure 5.
Induced current in the tissue.

in the tissue and adds to the resistance of the transmit coil. Additionally, there is some amount of inductance associated with this current flow.

The effect of the current flow in the tissue may be crudely approximated by a shorted inductance. The Kirchhoff's laws are:

$$
\begin{aligned}
(R + j\omega L)I + j\omega M_{tissue} I_{tissue} &= V \\
j\omega M_{tissue} I + (R_{tissue} + j\omega L_{tissue}) I_{tissue} &= 0
\end{aligned}
\tag{31}
$$

where M_{tissue} is the mutual inductance between the transmit coil and the shorted inductance.

Solving this for impedance $Z = V/I$:

$$
Z = R + \frac{\omega^2 M_{tissue}^2 R_{tissue}}{R_{tissue}^2 + \omega^2 L_{tissue}^2} + j\omega \left(L - \frac{\omega^2 M_{tissue}^2 L_{tissue}}{R_{tissue}^2 + \omega^2 L_{tissue}^2} \right)
\tag{32}
$$

The presence of the $\mathbf{E}_{current}$ electric field results in an increase of the resistance and a decrease of the inductance in the presence of the tissue. Generally, Eq. (32) can be written as:

$$
Z = R + R_{eddy} + j\omega (L - L_{eddy})
\tag{33}
$$

where the definitions of R_{eddy} and L_{eddy} can be inferred from the Eq. (32).

It can be observed that the tissue loading the coil leads to induced (eddy) currents in the tissue which causes power loss. This power loss in the tissue exhibits itself as an increased resistance and a decreased reactance of the transmit coil.

6.2 E_{charge} electric field

When we excite the transmit coil with voltage, there are electric charges that accumulate on the wiring near the coil terminals. When the coil is in close proximity to lossy tissue it can be modelled as a lossy dielectric between the plates of a parallel plate capacitor, as shown in **Figure 6**.

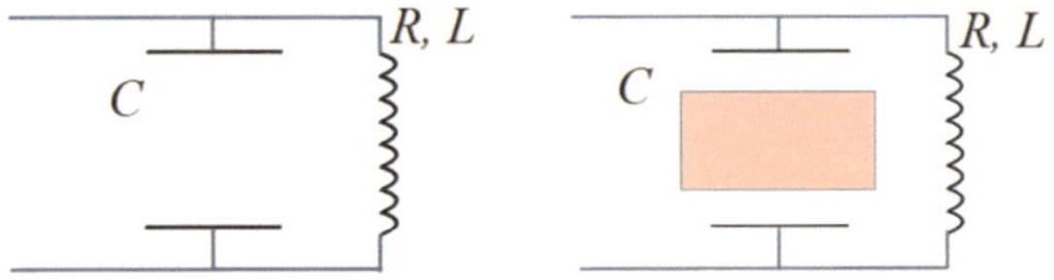

Figure 6.
Approaching tissue to the coil.

The Ampere's law is:

$$\nabla \times \mathbf{H} = \mathbf{J} + j\omega \mathbf{D} \tag{34}$$

where **H** is the magnetic field, **J** is the current density, **D** is the electric displacement. Taking divergence on both sides of (34) we get:

$$\mathrm{div}((\sigma_{tissue} + j\omega\varepsilon_0\varepsilon_{r,tissue})\mathbf{E}) = 0 \tag{35}$$

where σ_{tissue} is conductivity of tissue in S/m, $\varepsilon_{r,tissue}$ is relative electric permittivity of tissue in F/m, ε_0 is vacuum permittivity.

The normal component of the vector $(\sigma_{tissue} + j\omega\varepsilon_0\varepsilon_{r,tissue})E$ is preserved in the lossy tissue as shown in **Figure 7**.

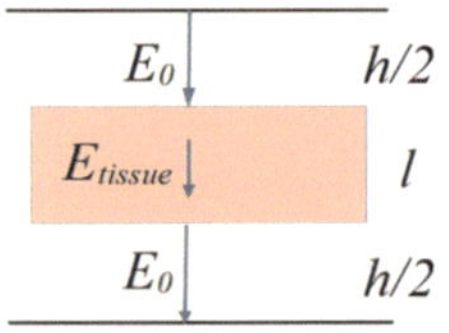

Figure 7.
Electric field inside the capacitor.

The electric fields inside the capacitor and outside of the tissue are related as follows:

$$E_{tissue} = \frac{E_0}{\varepsilon_{r,tissue} + \frac{\sigma_{tissue}}{j\omega\varepsilon_0}} \tag{36}$$

where E_0 is the electric field in air.

We denote the thickness of the tissue as l and the remaining free space between the plates of the capacitor as h. Voltage across the capacitor plates is:

$$V = hE_0 + lE_{tissue} = E_0\left(h + \frac{l}{\varepsilon_{r,tissue} + \frac{\sigma_{tissue}}{j\omega\varepsilon_0}}\right) \tag{37}$$

The electrical field in empty space between capacitor plates is:

$$E_0 = \frac{Q}{\varepsilon_0 A} \tag{38}$$

where Q is the charge on the capacitor plates and A is the area of the capacitor plates. Capacitance is:

$$C = \frac{Q}{V} = \frac{\varepsilon_0 A}{h + \frac{l}{\varepsilon_{r,tissue} + \frac{\sigma_{tissue}}{j\omega\varepsilon_0}}} \tag{39}$$

We note that the capacitance has an imaginary component. The impedance associated with this capacitance is calculated as:

$$Z_c = \frac{1}{j\omega C} = \frac{1}{j\omega\varepsilon_0 A}\left(h + \frac{l}{\varepsilon_{r,tissue} + \frac{\sigma_{tissue}}{j\omega\varepsilon_0}}\right) \tag{40}$$

This can be elaborated as:

$$Z_c = \frac{1}{j\omega\varepsilon_0 A}\left(h + \frac{l}{\varepsilon_{r,tissue}}\frac{1}{1 + \frac{\sigma_{tissue}{}^2}{\omega^2\varepsilon_0^2\varepsilon_{r,tissue}^2}}\right) + \frac{l}{A\sigma_{tissue}}\frac{1}{1 + \frac{\omega^2\varepsilon_0^2\varepsilon_{r,tissue}^2}{\sigma_{tissue}{}^2}} = \frac{1}{j\omega C_p} + R_p \tag{41}$$

where C_p and R_p are the effective capacitance and resistance and take the form:

$$C_p = \frac{\varepsilon_0 A}{h + \frac{l}{\varepsilon_{r,tissue}} \cdot \frac{\omega^2\varepsilon_0^2\varepsilon_{r,tissue}^2}{\sigma_{tissue}{}^2 + \omega^2\varepsilon_0^2\varepsilon_{r,tissue}^2}} \tag{42}$$

$$R_p = \frac{l}{A\sigma_{tissue}} \cdot \frac{\sigma_{tissue}{}^2}{\sigma_{tissue}{}^2 + \omega^2\varepsilon_0^2\varepsilon_{r,tissue}^2} \tag{43}$$

As we see from these formulas, the presence of the tissue between the capacitor plates leads to an increase of the effective capacitance C_p and the appearance of the effective resistance R_p. In the absence of tissue $R_p = 0$. When the coil is closer to a lossy dielectric medium like body tissue (e.g. muscle), we observe that the resonance frequency of the coil drops (detuning) and the ohmic losses increase.

To determine the resistance and reactance of a coil in close proximity to lossy tissue we develop an equivalent circuit shown in **Figure 8**. The impedance of the circuit in **Figure 8** is:

$$Z = \frac{1}{j\omega C_p + \frac{1}{R + j\omega L}} \tag{44}$$

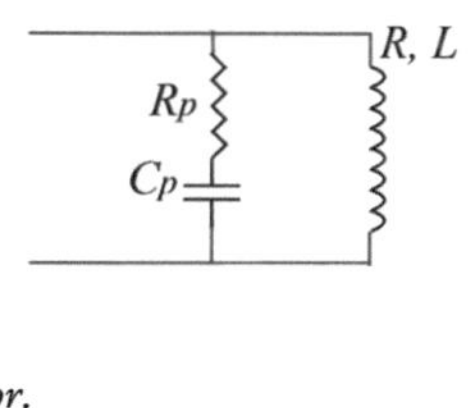

Figure 8.
Coil model with shunt capacitor and resistor.

We assume that the capacitive reactance $\frac{1}{\omega C_p}$ far exceeds the resistance R_p, and we neglect the resistance R_p. To further simplify this expression, we assume that the quality factor of the coil is much higher than unity ($R \ll j\omega L$). The self-resonance frequency of the coil is defined as:

$$\omega_s = \frac{1}{\sqrt{LC_p}} \tag{45}$$

We assume that the capacitive reactance $1/\omega C_p$ of the coil is much higher than the inductive reactance ωL. This implies that the Self Resonance Frequency (SRF) of the coil is much higher than the operating frequency ($\omega \ll \omega_s$), which is considered favourable for most practical coil designs.

With the aforementioned assumptions, the impedance of the coil simplifies to:

$$Z \approx \frac{R}{\left(1 - \frac{\omega^2}{\omega_s^2}\right)^2} + j\omega \frac{L}{1 - \frac{\omega^2}{\omega_s^2}} \tag{46}$$

From the above equation it can be observed that the proximity of lossy dielectric tissue results in an increase of the parasitic capacitance C_p and lowers the SRF of the coil ω_s due to an appearance of the parasitic resistance in series with the parasitic capacitance. This always results in an increase in the resistance of the coil. Depending on the coil geometry, dielectric properties of tissue near the coil and frequency of operation, the reactance may either decrease or increase when the coil is near lossy tissue.

7. Segmentation

Segmentation is a process of inserting additional capacitors in between the coil windings (see **Figure 9**). The capacitor placement is roughly equidistant throughout the windings of the coil. The purpose of the segmentation capacitors is to decrease the voltages between the terminals of the coil and between the turns of the coil.

The values of the segmentation capacitors are chosen to significantly decrease the visible inductance of the coil. There is no exact formula for the values of the segmentation capacitors, but our recommendation is as follows:

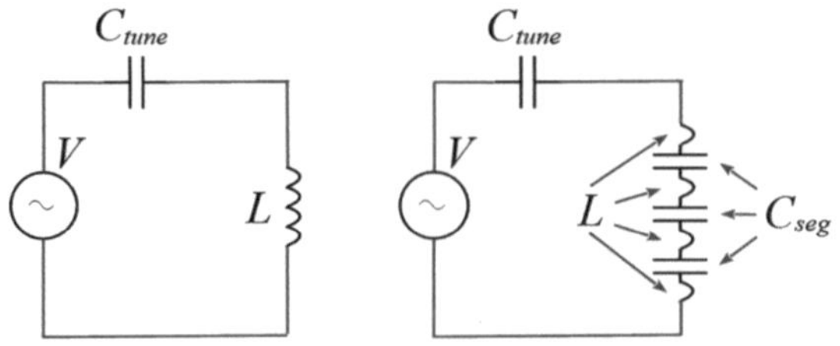

Figure 9.
Schematic of non-segmented and segmented loaded coils.

$$C_{seg} = \frac{N}{\omega^2 L} \tag{47}$$

where N is the number of segments and L is the coil inductance. If we have N segments, then we have N-1 segmentation capacitors. The cumulative effect of N-1 segmentation capacitors placed in series is represented as the cumulative segmentation capacitance:

$$C_{seg,c} = \frac{N}{(N-1)\omega^2 L} \tag{48}$$

7.1 Effect of segmentation on the coil resistance and inductance

Figure 10 shows the equivalent circuit of a non-segmented and segmented coil when the coil is loaded by body tissue. The segmentation affects the coil impedance by reducing the electric charges on the wiring of the coil. Mathematically, the effect of segmentation capacitors can be introduced by modifying Eq. (46) as follows:

$$Z \approx \frac{R + R_{eddy}}{\left(1 - \frac{\omega^2}{\omega_s^2}\left(1 - \frac{1}{\omega^2 L C_{seg,c}}\right)\right)^2} + j\omega\left(\frac{L\left(1 - \frac{1}{\omega^2 L C_{seg,c}}\right)}{1 - \frac{\omega^2}{\omega_s^2}\left(1 - \frac{1}{\omega^2 L C_{seg,c}}\right)} - L_{eddy}\right) \tag{49}$$

The SRF ω_s depends on whether the coil is loaded or not: loaded value $\omega_{s,loaded}$ is smaller than the unloaded value $\omega_{s,unloaded}$. We consider the ratio $\omega^2/\omega_{s,loaded}^2$ much less than unity, otherwise the coil would not be functioning correctly.

We will now study the effect of tissue loading on both the non-segmented and the segmented coils. For the non-segmented coil, the expression $\left(1 - \frac{1}{\omega^2 L C_{seg,c}}\right)$ is unity. The coil is tuned under unloaded condition by placing a tuning capacitor C_{tune} in series with it. So, the reactance of the unloaded coil is zero.

For the tuned non-segmented coil, the unloaded and loaded impedances are:

$$Z_{\substack{unloaded \\ non-seg}} \approx \frac{R}{\left(1 - \frac{\omega^2}{\omega_{s,unloaded}^2}\right)^2} + j\omega\left(\frac{L}{1 - \frac{\omega^2}{\omega_{s,unloaded}^2}} - \frac{1}{\omega^2 C_{tune}}\right) \tag{50}$$

$$Z_{\substack{loaded \\ non-seg}} \approx \frac{R + R_{eddy}}{\left(1 - \frac{\omega^2}{\omega_{s,loaded}^2}\right)^2} + j\omega\left(\frac{L}{1 - \frac{\omega^2}{\omega_{s,loaded}^2}} - L_{eddy} - \frac{1}{\omega^2 C_{tune}}\right) \tag{51}$$

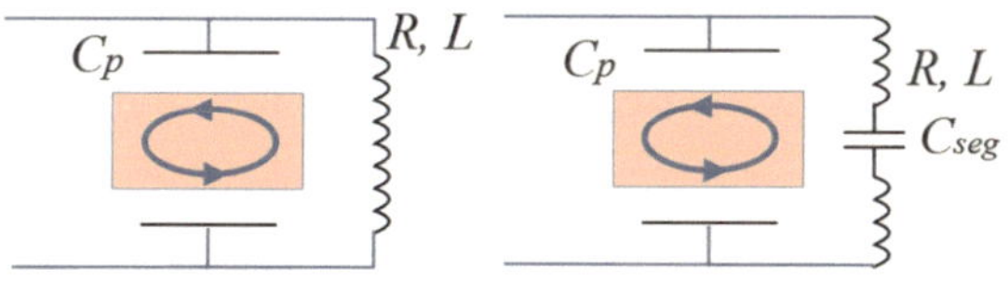

Figure 10.
Non-segmented and segmented loaded coils.

The difference between these values is:

$$Z_{\substack{loaded \\ non-seg}} - Z_{\substack{unloaded \\ non-seg}} \approx R\left(\frac{2\omega^2}{\omega_{s,loaded}^2} - \frac{2\omega^2}{\omega_{s,unloaded}^2}\right) + R_{eddy}\left(1 + \frac{2\omega^2}{\omega_{s,loaded}^2}\right) + j\omega L\left(\frac{\omega^2}{\omega_{s,loaded}^2} - \frac{\omega^2}{\omega_{s,unloaded}^2}\right) \quad (52)$$

For the segmented coil, the expression $\left(1 - \frac{1}{\omega^2 L C_{seg,c}}\right)$ simplifies to $1/N$ for a coil with N segments. Again, we tune the coil when it is not loaded, so the reactance of the unloaded coil is zero.

For the tuned segmented coil, the unloaded and loaded impedances are:

$$Z_{\substack{unloaded \\ seg}} \approx \frac{R}{\left(1 - \frac{\omega^2}{\omega_{s,unloaded}^2} \cdot \frac{1}{N}\right)^2} + j\omega\left(\frac{\frac{1}{N} \cdot L}{1 - \frac{\omega^2}{\omega_{s,unloaded}^2} \cdot \frac{1}{N}} - \frac{1}{\omega^2 C_{tune}}\right) \quad (53)$$

$$Z_{\substack{loaded \\ seg}} \approx \frac{R + R_{eddy}}{\left(1 - \frac{\omega^2}{\omega_{s,loaded}^2} \cdot \frac{1}{N}\right)^2} + j\omega\left(\frac{\frac{1}{N} \cdot L}{1 - \frac{\omega^2}{\omega_{s,loaded}^2} \cdot \frac{1}{N}} - L_{eddy} - \frac{1}{\omega^2 C_{tune}}\right) \quad (54)$$

The difference between these values is:

$$Z_{\substack{loaded \\ seg}} - Z_{\substack{unloaded \\ seg}} \approx \frac{R}{N}\left(\frac{2\omega^2}{\omega_{s,loaded}^2} - \frac{2\omega^2}{\omega_{s,unloaded}^2}\right) + R_{eddy}\left(1 + \frac{2\omega^2}{N\omega_{s,loaded}^2}\right) + j\omega\frac{L}{N^2}\left(\frac{\omega^2}{\omega_{s,loaded}^2} - \frac{\omega^2}{\omega_{s,unloaded}^2}\right) \quad (55)$$

Comparing Eqs. (52) and (55) we observe that for a segmented coil: (a) the resistance increase due to proximity of lossy tissue is lower than that for unsegmented coil, (b) the reactance increase due to the proximity of lossy tissue is lower than that for unsegmented coil. This is clearly due to the $1/N$ and$1/N^2$ factors responsible for this effect. Therefore, segmenting the coil significantly improves the robustness of the coil to the deleterious effects of the lossy body tissue.

7.2 Effect of segmentation on the transmit coil figure of merit

Figure 11 shows the equivalent circuit for non-segmented and segmented loaded transmit coils.

The input current splits into two branches: current I that flows through the ideal inductor L and parasitic current I_p that flows through the capacitor C_p and resistor R_p

These two currents are related as follows:

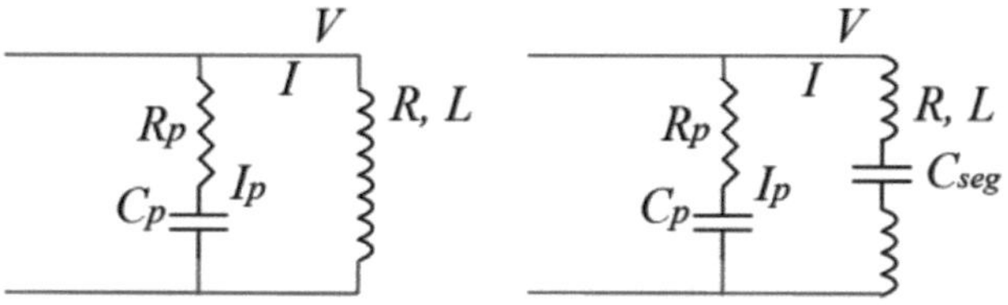

Figure 11.
Non-segmented and segmented loaded transmit coils.

$$I_p \approx -\frac{\omega^2}{\omega_s^2}\left(1-\frac{1}{\omega^2 L C_{seg,c}}\right) I \tag{56}$$

where we neglected the resistances R and R_p.

In the transmit figure of merit ($\frac{B_t}{\sqrt{2P_{in}}}$), the magnetic flux density B_t depends on the current I in the coil. If we keep the current I fixed, then the magnetic flux density will also remain fixed.

The power needed to generate the current I (and magnetic flux density B_t) is:

$$\begin{aligned} P_{in} &= \frac{1}{2}(R+R_{eddy})I^2 + \frac{1}{2}R_p I_p^2 \\ &= \frac{1}{2}(R+R_{eddy})I^2 + \frac{1}{2}R_p \frac{\omega^4}{\omega_s^4}\left(1-\frac{1}{\omega^2 L C_{seg,c}}\right)^2 I^2 \end{aligned} \tag{57}$$

The figure of merit is then:

$$F_t = \frac{B_t}{\sqrt{2P_{in}}} = \frac{B_t/I}{\sqrt{(R+R_{eddy}) + R_p \frac{\omega^4}{\omega_s^4}\left(1-\frac{1}{\omega^2 L C_{seg,c}}\right)^2}} \tag{58}$$

We observe that segmentation leads to an increase in the transmit figure of merit of a coil. This is because for the segmented coil, the voltage V across the terminals of the coil is reduced by $\frac{I}{j\omega C_{seg,c}}$. This means that the current through the parasitic resistance R_p will be less and, therefore, the corresponding ohmic loss will be less.

7.3 Effect of segmentation on the receive figure of merit

Figure 12 shows the equivalent circuit for non-segmented and segmented loaded receive coils. In the figure, A_r is the effective aperture area of the receive coil, B_t is the incident magnetic field from the transmit coil and $\omega B_t A_r$ is the voltage appearing across the receive coil terminals.

For the non-segmented coil, the optimal loaded resistance is:

$$R_L \approx \frac{R_r + R_{eddy}}{\left(1-\frac{\omega^2}{\omega_s^2}\left(1-\frac{1}{\omega^2 L C_{seg,c}}\right)\right)^2} \tag{59}$$

Currents through the voltage source $\omega B_t A_r$ and through the load R_L

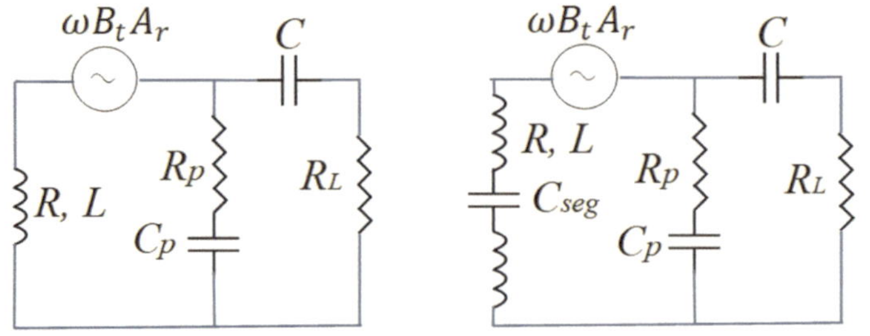

Figure 12.
Non-segmented and segmented loaded receive coils.

$$I_r \approx \frac{1}{2} \cdot \frac{\omega B_t A_r}{R_r + R_{eddy}} \tag{60}$$

$$I_L \approx \frac{1}{2} \cdot \frac{\omega B_t A_r}{R_r + R_{eddy}} \cdot \left(1 - \frac{\omega^2}{\omega_s^2}\left(1 - \frac{1}{\omega^2 L C_{seg,c}}\right)\right) \tag{61}$$

The delivered power is:

$$P_L \approx \frac{(\omega B_t A_r)^2}{8(R_r + R_{eddy})} \cdot \left(1 - \frac{R_p}{R_r + R_{eddy}} \frac{\omega^4}{\omega_s^4}\left(1 - \frac{1}{\omega^2 L C_{seg,c}}\right)^2\right) \tag{62}$$

Again, we observe that the receive figure of merit $\sqrt{8P_L}/\omega B_t$ without the segmentation capacitor ($C_{seg,c} \to inf$) is lower than the receive figure of merit with segmentation capacitor. Segmentation, therefore, leads to the increase of the receive figure of merit.

8. Full wave simulations and measurements

To verify the theory presented in the previous sections, a PCB spiral coil is modelled in Ansys HFSS, as shown in **Figure 13**. A trace on the bottom layer is used to connect the inner terminal of the coil through a via, to form a closed loop. The

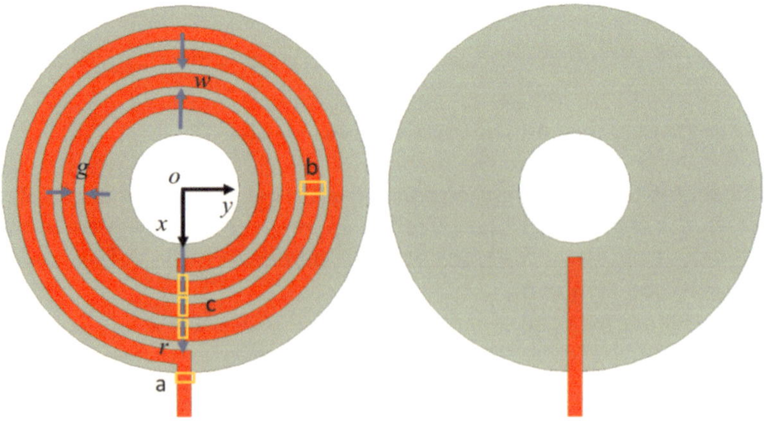

Figure 13.
Top (left) and bottom (right) view of the spiral coil. (a is the tuning capacitor, b is the capacitor for 2 segments and c is the capacitors for 4 segments).

Radius R, mm	Trace width w, mm	Inter-trace distance g, mm	No. of turns
35	3	2	4

Table 1.
Coil dimensions.

locations of the segmentation capacitors are indicated for different segmentation numbers. The dimensions of the coil are listed in **Table 1**. The substrate is a 1.5 mm FR-4 with 1 oz. copper.

The coil is firstly simulated without any capacitors. The inductance of the coil can be obtained as:

$$L = \frac{\mathrm{im}(Z_{11})}{\omega} \tag{63}$$

where Z_{11} is the input impedance of the coil, and ω is the radian frequency. The tuning capacitor can be calculated as:

$$C_{tune} = \frac{1}{\omega\,\mathrm{im}(Z_{11})} \tag{64}$$

The values of the segmentation capacitors are calculated using Eq. (47). In practice, the values of the segmentation capacitors would be a little higher due to the parasitic capacitance of the coil itself.

Once the coil is tuned to resonate at the desired frequency, either with or without segmentation, the resistance can be obtained as:

$$R = \mathrm{re}(Z_{11}) \tag{65}$$

To evaluate the effect of the segmentation on the resistance, three cases are compared by simulation and verified with measurement: (a) coil without segmentation with one series capacitor to resonate the coil; (b) coil with one segmentation capacitor splitting the coil wiring into two equal segments; (c) coil with three segmentation capacitors splitting the coil wiring into four equal segments.

8.1 Coil resistance

The fabricated coils with and without segmentation are shown in **Figure 14**. All the coils are tuned to resonate at 27.12 MHz. A comparison of the simulated and the measured resistance with and without segmentation is shown in **Table 2**. Excellent

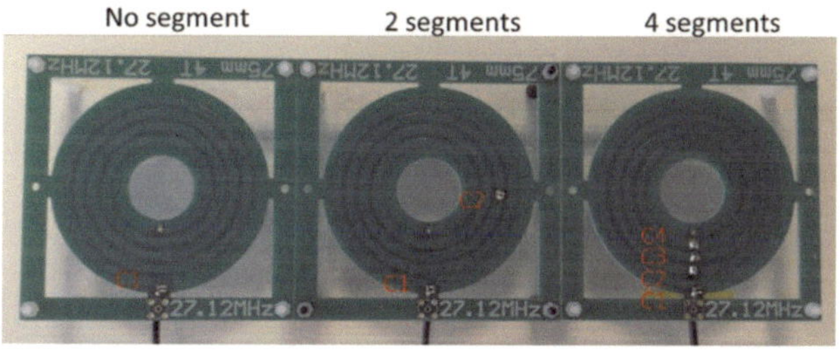

Figure 14.
Fabricated coils with and without segmentation capacitors.

Coil	Simulated/Measured	Capacitor values (pF)	Resistance (Ω)
Coil non-segmented	Simulated	C1 = 30	0.69
	Measured	C1 = 35	0.86
Coil with 2 segments	Simulated	C1 = C2 = 64	0.50
	Measured	C1 = C2 = 68	0.53
Coil with 4 segments	Simulated	C1 = C2 = C3 = C4 = 134	0.39
	Measured	C1 = C2 = C3 = C4 = 139	0.42

Table 2.
Simulated and measured resistance for segmented and non-segmented coils.

agreement is found between the simulations and the measurements. The values of the capacitors needed to resonate the coil at 27.12 MHz are higher than the values calculated using Eq. (64). The measured resistances of the Printed Circuit Board (PCB) coils are higher than the simulated ones because of the extra capacitance and loss from the testing cable and connector which is not included in the simulations. What is clear from both simulation and measurement is that the addition of segmentation capacitors significantly reduces the coil resistance and the associated power loss in the coil.

8.2 Figure of merit F_t

The electromagnetic (EM) fields generated from the coils can be simulated in HFSS. The FoM F_t is used to compare the coils with and without segmentation. For ease of comparison, the fields along the X, Y, Z directions are plotted, where the origin of the coordinate system is the center of the coil, and the coil is placed at the XY-plane, as indicated in **Figure 13**. The electric fields normalised by the input power as $\frac{E}{\sqrt{P_{in}}}$ are also plotted. From **Figure 15** we can observe the following trends:

1. Segmentation can increase the magnetic field strength without changing the field distribution generated by the coil. So, the designer can start with the non-segmented coil to optimise the field coverage first.

2. Segmentation can suppress the electric field in the direction perpendicular to the coil surface (moving away from the coil).

3. Segmentation can significantly alter the electric field distribution near the coil surface with localised maxima close to the capacitors.

Figure 16 plots the heat map of the magnitude of the electric field in the PCB substrate indicating that, as the electric fields are concentrated around the segmentation capacitors, the dielectric loss in the substrate is reduced.

8.3 Transfer efficiency

In this section, the effect of the segmentation on the power transfer efficiency is evaluated in both simulations and measurements. The receive coil shares the same HFSS model as the transmit coil, only with different dimensions and number of turns.

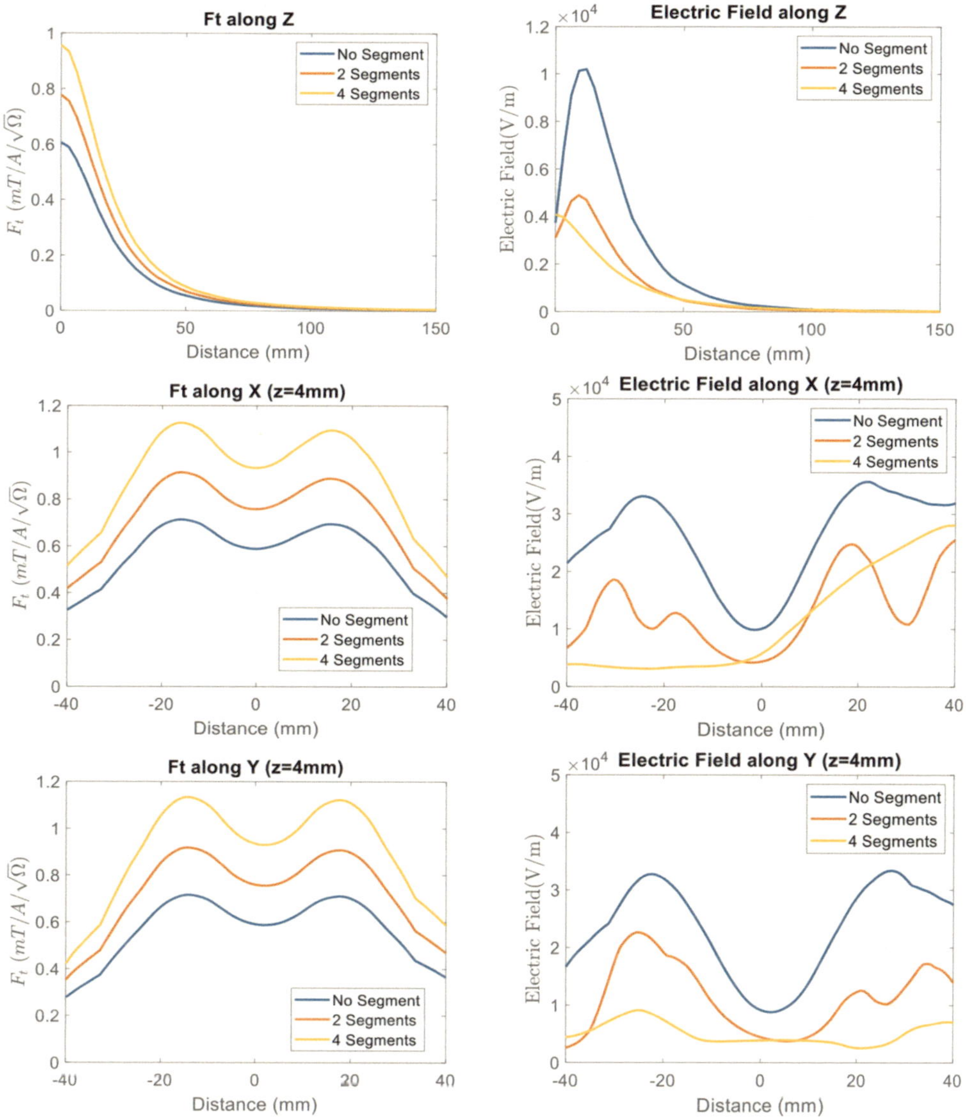

Figure 15.
FoM Ft and normalised electric field plots along Z, X, and Y directions. The plots for X and Y directions are at z = 4 mm height.

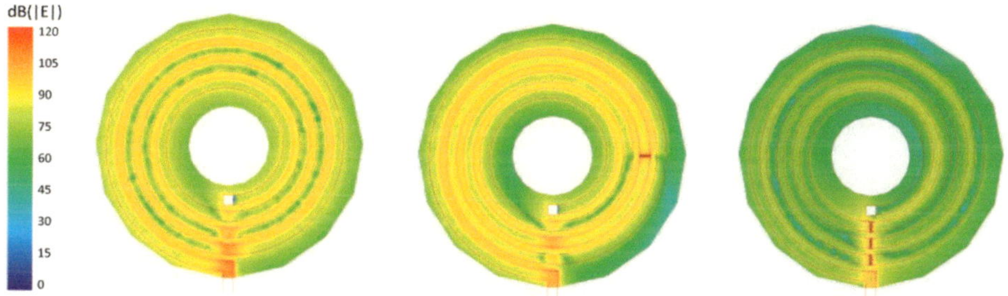

Figure 16.
Electric field in the substrate. (left: no segment; middle: 2 segments; right: 4 segments).

Radius R, mm	Trace width w, mm	Inter-trace distance g, mm	No. of turns
4.88	0.24	0.26	5

Table 3.
Geometry of the receive coil.

The parameters of the receive coil that is simulated is shown in **Table 3**. The substrate is 0.8 mm FR-4 with 1 oz. copper.

The receive coil is placed 10 mm above the transmit coil with its center aligned with the center of the transmit coil, as shown in **Figure 17**. To investigate the loading effect of the human body, a hand is placed close to the coil. To measure the transfer efficiency, we perform the following steps:

a. Calibrate two ports of network analyser at the frequency of interest.

b. Tune the transmit coil with series capacitor, connect it to the network analyser and measure its resistance. Repeat for the receive coil.

c. Place the two coils in proximity of each other and connect them to the network analyser.

d. Measure and save the S-matrix of the system (2×2 matrix).

e. Convert S-matrix to Z-matrix using:

$$Z = Z_0(U + S)(U - S)^{-1} \tag{66}$$

where U is the unity matrix, Z_0=50 Ω

f. Calculate mutual inductance using:

$$M = \frac{1}{\omega}\text{im}(Z_{21}) \tag{67}$$

g. Calculate the mutual quality factor using Eq. (10).

h. Calculate the efficiency using Eqs. (14) or (18), depending on if we choose to maximise delivered power or transfer efficiency.

Figure 17.
Measurement setup for the transfer efficiency (left: top view; middle: side view; right: a hand is close to the coils).

While designing a WPT system for medical implants, care must be taken to understand the various use cases and user interactions and its implications on power delivery. An important decision that needs be made is whether a design is maximised for delivered power to an implant or efficiency of the WPT link. On one hand, if it is challenging for the receive coil inside of the implant to harvest the needed amount of power, then maximising the delivered power is preferential. On the other hand, most body worn charging systems are battery-powered and have a limited amount of available power to deliver to the implant. So, maximising the efficiency directly results in longer duration before the battery runs out on the charger and needs to be recharged by the patient or the caregiver.

As an example for this article, we chose to calculate the transfer efficiency using Eq. (18) for the coils with and without segmentation. The transfer efficiency is also simulated in HFSS for comparison. Furthermore, a 200 mm × 200 mm 3-layered tissue stack model is placed 2 mm above the transmit coil and the receive coil is embedded in the fat layer with the same 10 mm distance to the transmit coil in HFSS, as shown in **Figure 18**. The thickness of the skin, fat and muscle is 2 mm, 23 mm and 20 mm, respectively.

The simulated and measured transfer efficiencies are summarised in **Table 4**. We observe that the transfer efficiency both in air and in tissue can be improved with segmentation. Although we have done the calibration to minimise the effect of the cables and connectors, the measured efficiency is still a little lower than the simulated one, which is not surprising. However, with segmentation, we can see that the measured efficiency is much closer to the simulated one. It implies that the segmentation can reduce the loading effect of the environment (e.g. cables). The measured efficiency of a coil without segmentation in the presence of body tissue (hand) shows a significant drop from 61.9% to 46.2%, while the measured efficiency of a coil with two and four segmentations shows only a drop from 65.0% to 62.3% and 67.2% to 66.5%,

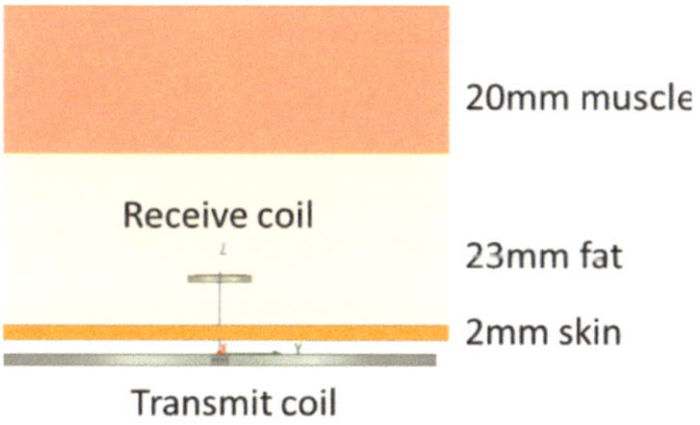

Figure 18.
Transfer efficiency simulation with tissue stack.

Segmentation	no segments	2 segments	4 segments
Simulated in air	67.4	68.6	70.7
Simulated with tissue stack	36.3	41.6	42.8
Measured in air	61.9	65.0	67.2
Measured with a hand nearby	46.2	62.3	66.5

Table 4.
Transfer Efficiency (%).

respectively. This clearly indicates that the segmented coils are more robust to the presence of lossy tissue. In case of the simulated coils, the tissue of **Figure 18** has a much larger effect on the coil, because there is large drop in efficiency when the tissue is nearby for non-segmented and segmented coils.

8.4 Figure of merit $F_{t,SAR}$

The SAR value in tissue is simulated in HFSS using the same tissue model as in the previous section. For a fair comparison, the SAR is also normalised by the input power as:

$$\overline{SAR} \equiv \frac{SAR}{P_{in}} \tag{68}$$

Figure 19 compares the distribution of the peak average SAR where the SAR value has been normalised to the peak SAR value for each of the three coil designs presented. The IEC/IEEE 62704-4 method is used to calculate the peak average 1 g SAR. Without segmentation, the regions of high SAR value occur at the overlapping area between the trace on the top and bottom layers of the PCB. This is because there is high stored electric field between the layers resulting in high parasitic capacitance. For the coil with two segments, the regions of high SAR value are between the segmentation capacitor and the areas of overlap between the top and the bottom layers of the PCB. Both these regions have high parasitic capacitance. For the coil with four segments, the 3 segmentation capacitors are lined close to each other resulting in a region of high stored electric field. This results in the coil with four segments having higher peak average SAR compared to the coil with two segments, but still lower than the coil with no segmentation capacitors. The results also clearly indicate that the locations of the segmentation capacitors play a critical role in reducing the peak average SAR.

Another important advantage of introducing segmentation capacitors in the wiring of the coils (or along the coil traces) is that the distribution of the averaged SAR and its maximum value can be significantly altered by optimising the locations of the segmentation capacitors along the coil. For example, the 4-segment coil (in **Figure 14**) has the three segmentation capacitors in close proximity, all in the same sector of the circular coil. For the same coil **Figure 20** shows a significantly different SAR distribution and reduced maximum SAR value when the three segmentation capacitors are

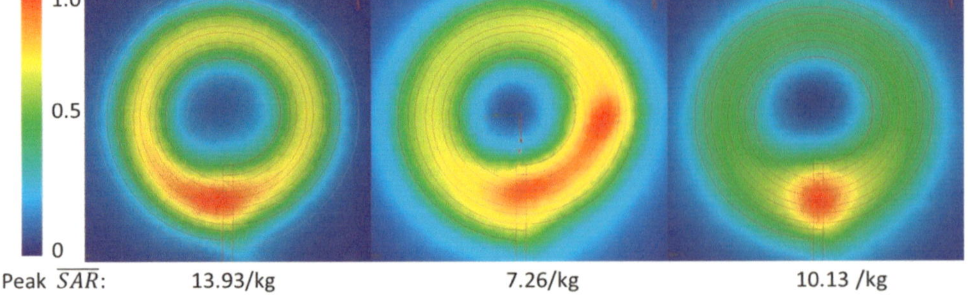

Figure 19.
The SAR maps in the tissue. (top: coil with no segments; middle: coil with two segments; bottom: coil with four segments).

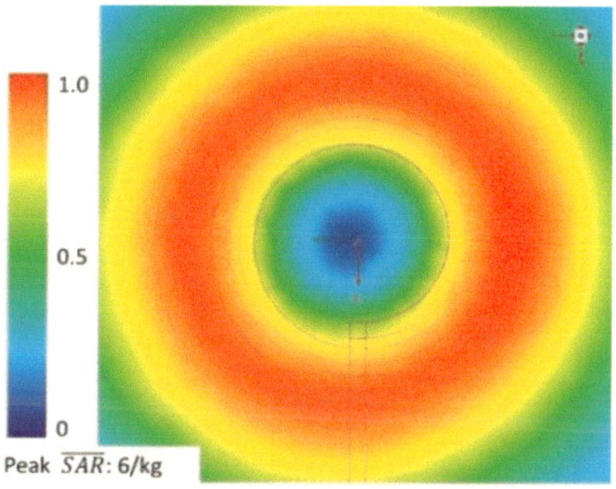

Figure 20.
SAR map of the coil with spread segmentation capacitors.

spread along the coil with 90-degree separation. The heat map shows that the regions of high SAR value are shaped like a circle and the peak value of the SAR is reduced by 40%. It should be noted that the tuning capacitor is not shown in this plot because it is placed at the far end of the coil input.

With the normalised SAR and the coil resistance, the maximum allowed current within FCC limit can be calculated as

$$I_{max} = \sqrt{2P_{max}/R_{loaded}} \tag{69}$$

where $P_{max} = 1.6/\overline{SAR}$, and R_{loaded} is the coil resistance when the coil is in close proximity to lossy tissue.

Table 5 summarises the coil impedance, SAR, $F_{t,SAR}$ and maximum current compliant to FCC limit with different segmentations. The coil without segmentation is also listed for comparison. It is noted that when the segmentation capacitors are spread along the coil, the peak SAR is significantly reduced, and the maximum current within FCC limit is increased. From the **Table 5** we observe that coil resistance decreases as

Coil	**Non-segmented**	**Segmented with 2 segments**	**Segmented with 4 segments**	**Segmented with 4 segments, spread**
Z_{11} (Ω) in air	0.69 + j1.27	0.50 + j0.94	0.39 + j4.69	0.42 + j4.28
Z_{11} (Ω) with tissue	1.88 + j9.65	1.13 + j2.17	0.97 + j5.51	0.94 + j4.40
$\overline{SAR}$ (1/kg)	13.93	7.26	10.13	6.00
$F_t\left(\text{mT}/\left(\text{A}\sqrt{\Omega}\right)\right)$ with tissue (z = 10 mm)	0.0938	0.1015	0.1089	0.1070
$F_{t,SAR}$ $\left(\text{mT}/\left(\sqrt{\text{W/kg}}\right)\right)$ (z = 10 mm)	0.0251	0.0377	0.0342	0.0437
Max allowed current within FCC 1.6 W/kg limit (mA)	349.6	624.6	570.7	753.2

Table 5.
Comparison of coil impedance, SAR,F_t,$F_{t,SAR}$ and the max current compliant to FCC limit with segmentations.

Coil	Non-segmented	Segmented with 2 segments	Segmented with 4 segments	Segmented with 4 segments, spread
ΔR, (Ω)	1.19	0.63 (−47%)	0.58 (−51%)	0.52 (−56%)

Table 6.
Change in resistance.

Coil	Non-segmented	Segmented with 2 segments	Segmented with 4 segments	Segmented with 4 segments, spread
ΔX, (Ω)	8.38	1.23 (−85%)	0.82 (−90%)	0.12 (−98.6%)
$\frac{\Delta X_{non-segmented}}{\Delta X_{segmented}}$	1	6.8	10.2	69.8
N^2 factor	1	4	16	16

Table 7.
Comparing the change in reactance with the predicted one.

we increase coil segmentation: from 0.69 Ω to 0.39 Ω if in air (−44%) and from 1.88 Ω to 0.94 Ω if near tissue (−50%).

When we compare the "Z_{11} in air" and "Z_{11} with tissue", we observe that both resistance (real part of impedance) and reactance (imaginary part of impedance) increase when the tissue is in the proximity of the transmit coil. For example, the resistance grows from 0.69 Ω to 1.88 Ω (+172%) for the non-segmented coil when the tissue is approached.

The figure of merit F_t grows by 16% and $F_{t,SAR}$ grows by 74% as we increase coil segmentation.

Let us now compare the change in resistance ΔR and the change in reactance ΔX for the four coils as shown in **Table 6**. We observe that change in resistance ΔR is decreasing with the progressing segmentation, up to 56%. In the following table we show the measured change in reactance ΔX.

From the **Table 7** we observe that there is a decrease in the change in reactance ΔX as the number of segments in the transmit coil, N increases. Assuming that Leddy contribution is negligible, the Eq. (55) predicts that the ratio of non-segmented ΔX to segmented ΔX would grow as N^2 . For the "4-segment spread" the ratio of changes in reactance far exceeds the prediction of N^2. Spreading the segmentation capacitors away from one another, significantly helps to stabilise the transmit coil. While the numbers do not exactly match, the trend showing the increase in the ratio is as predicted in Eq. (55).

9. Conclusions

In this work, we introduced and derived unique optimisation metrics for designing efficient transmit and receive coils for magnetics based WPT solutions for medical implants. We reviewed the regulations imposed on WPT systems for medical implants in the US and EU regions and determined the most limiting parameters that place a bound on the maximum current that can be driven into a coil. We derived the expressions for delivered power and efficiency considering the identified regulatory limits for the transmit and the receive coil currents. We demonstrated that, under certain conditions, the system figure of merit can be "split" into transmit figure of merit and receive figure of merit permitting independent evaluation of transmit and receive coils.

We studied the effect of lossy tissue on the performance of transmit coils from a circuit theory perspective. We showed that the resistance of the transmit coil increases in the presence of tissue because of two types of electromagnetic phenomena: (i) increase in parasitic capacitance between the opposite charges accumulating in the surfaces of the coil (charge contribution); (ii) the eddy currents in the tissue (current distribution). We showed that the change in reactance of the coil due to the presence of lossy tissue is dependent on which contribution (charge or current) is more significant.

With this improved understanding of the effect of lossy tissue on coils we introduced the concept of segmented on-body transmit coils. We hypothesised that the resistance and reactance of a transmit coil with segmentation capacitors is less sensitive to the presence of lossy tissue. We derived the impact of segmentation on the transmit figure of merit and the receive figure of merit of a coil using circuit theory. We showed analytically that segmented coils have the potential to significantly improve both (transmit and receive) figures of merit, thereby positively affecting the efficiency of a WPT system.

To validate our hypothesis and assertions we built PCB coil prototypes at 27.12 MHz with and without segmentation. We performed full wave simulations using HFSS models of the same coils. We showed through simulations and measurements that the resistance of the transmit coil reduces substantially (as much as 50%) when we went from no segmentation to up to four segments (with three segmentation capacitors). We also confirmed that the proximity of lossy tissue has a significantly smaller effect on segmented transmit coil. We noted that, on the specific coils we built, we measured that the change in reactance of a coil between air and close proximity of tissue reduced from 4.2% (for non-segmented coil) to 0.06% (for segmented coil with capacitors uniformly spread). We also confirmed that the transmit figures of merit (F_t and $F_{t,SAR}$) of the segmented coil are higher than those of the non-segmented coil. F_t grew by 16% and $F_{t,SAR}$ grew by 74% as we increased the level of segmentation. We have found that the way the segmentation capacitors are spaced on the coil has a significant effect on coil performance and the distribution of electric field close to the wiring of the coil. This is an important result as the number of segmentation capacitors and their distribution to break up the coil wiring controls the distribution of electric field and will be very useful in controlling not just SAR but also to reduce coupling with the internal electronics of a charger.

Acknowledgements

We are grateful to Mark Norris, Richard Davies, Matthew Armean-Jones, and Olympia Karadima for their useful feedback that helped in writing this chapter.

Author details

Yun Tao[1*], Rosti Lemdiasov[2], Arun Venkatasubramanian[2] and Marshal Wong[1]

1 Cambridge Consultants, Singapore

2 Cambridge Consultants, Boston, MA, USA

*Address all correspondence to: yun.tao@cambridgeconsultants.com

References

[1] Oxley TJ, Opie NL, John SE, et al. Minimally invasive endovascular stent-electrode array for high-fidelity, chronic recordings of cortical neural activity. Nature Biotechnology. 2016;**34**:320-327

[2] Farra R, Sheppard NFJ, McCabe L, Neer RM, Anderson JM, Santini JTJ, et al. First-in-human testing of a wirelessly controlled drug delivery microchip. Science Translational Medicine. 2012; **4**(122):122ra21

[3] Ahmadi MM, Jullien GA. A wireless-implantable microsystem for continuous blood glucose monitoring. IEEE Transactions on Biomedical Circuits and Systems. 2009;**3**(3):169-180

[4] Abiri P, Abiri A, Packard RRS, Ding Y, Yousefi A, Ma J, et al. Inductively powered wireless pacing via a miniature pacemaker and remote stimulation control system. Scientific Reports. 2017;**7**(1):6180

[5] Li Q, Zhang Z, Mao J. Design of wireless power transfer with high efficiency for biomedical implants. Journal of Beijing Institute of Technology. 2022;**31**(1):53-60

[6] Sun G, Muneer B, Li Y, Zhu Q. Ultracompact implantable design with integrated wireless power transfer and RF transmission capabilities. IEEE Transactions on Biomedical Circuits and Systems. 2018;**12**:281-291

[7] Al-Kalbani AI, Yuce MR, Redouté J. A biosafety comparison between capacitive and inductive coupling in biomedical implants. IEEE Antenna Wireless Propagation Letters. 2014;**13**:1168-1171

[8] Jow UM, Ghovanloo M. Optimization of data coils in a multiband wireless link for neuroprosthetic implantable devices. IEEE Transactions on Biomedical Circuits and Systems. 2010;**4**(5):301-310

[9] Campi T, Cruciani S, De Santis V, Maradei F, Feliziani M. Near field wireless powering of deep medical implants. Energies. 2019;**12**(14):2720

[10] Haerinia M, Shadid R. Wireless power transfer approaches for medical implants: A review. Signals. 2020;**1**(2): 209-229

[11] Lemdiasov R, Venkatasubramanian A. Transmit coil design for wireless power transfer for medical implants. In: 2017 39th Annual International Conference of the IEEE Engineering in Medicine and Biology Society (EMBC). 2017. pp. 2158-2161

[12] Zhang J, Das R, Zhao J, Mirzai N, Mercer J, Heidari H. Battery-free and wireless technologies for cardiovascular implantable medical devices. Advanced Material technology. 2022:2101086

[13] Bao J, Hu S, Xie Z, Hu G, Lu Y, Zheng L. Optimization of the coupling coefficient of the inductive link for wireless power transfer to biomedical implants, Article. International Journal of Antennas and Propagation, ID. 2022; **8619514**:12

[14] Shan D, Wang H, Cao K, et al. Wireless power transfer system with enhanced efficiency by using frequency reconfigurable metamaterial. Science Report. 2022;**12**:331

[15] Stein ALF, Kyaw PA, Sullivan CR. Wireless power transfer utilizing a high-Q self-resonant structure. IEEE Transactions on Power Electronics. 2019; **34**(7):6722-6735. DOI: 10.1109/ TPEL.2018.2874878

[16] Fu M, Zhang T, Ma C, Zhu X. Efficiency and optimal loads analysis for multiple-receiver wireless power transfer systems. IEEE Transactions on Microwave Theory and Technologies. 2015;**63**(3):801-812

[17] Monti G, Mongiardo M, Minnaert B, Costanzo A, Tarricone L. Optimal terminations for a SIMO resonant inductive WPT link. Energies. 2020; **2020**(13):5157. DOI: 10.3390/en13195157

[18] Zargham M, Gulak PG. Maximum achievable efficiency in near-field coupled power transfer systems. IEEE Transaction on Biomedical Circuits and Systems. 2012;**6**(3):228-245. DOI: 10.1109/TBCAS.2011.2174794

[19] Minnaert B, Stevens N. Single variable expressions for the efficiency of a reciprocal power transfer system. International Journal of Circuit Theory and Applications. 2017;**45**:1418-1430

[20] Ohira T. Extended k-Q product formulas for capacitive and inductive coupling wireless transfer schemes. IEICE Electronics Express. 2014;**11**(9): 1-7. DOI: 10.1587/elex.11.20140147

[21] Ohira T. A Lucid Design Criterion for Wireless Power Transfer Systems to Enhance their Maximum Available Efficiency Sendai: Asia Pacific Microwave Conference (APMC). Sendai, Japan: IEEE; 2014

[22] Cho J, Sun J, Kim H, Fan J, Lu Y, Pan S. Coil design for 100 kHz and 6.78 MHz WPT system: Litz and solid wires and winding methods. IEEE Internationa l Symposium on Electromagnetic Compa tibility & Signal/Power Integrity (EMCSI). 2017:803-806. DOI: 10.1109/ ISEMC.2017.8077977

[23] Shinohara N. Power without wires. IEEE Microwave Magazine. 2011;**12**(7): S64-S73. DOI: 10.1109/MMM.2011. 942732

[24] Sharma PK. Concepts, quality factor and efficiency of wireless power transfer. International Journal of Engineering Research & Science (IJOER). 2015;**1**(1)

[25] Alderman DW, Grant DM. An efficient decoupler coil design which reduces heating in conductive samples in superconducting spectrometers. Journal of Magnetic Resonance. 1979;**36**(3): 447-451. DOI: 10.1016/0022-2364(79) 90123-9

[26] Roemer PB, Edelstein WA, Hayes CE, Souza SP, Mueller OM. The NMR Phased Array: Magnetic Resonance in Medicine. 1990;**16**(2):192-225. DOI: 10.1002/mrm.1910160203

[27] Mirbozorgi SA, Jia Y, Canales D, Ghovanloo M. A wirelessly-powered homecage with segmented copper foils and closed-loop power control. IEEE Transactions on Biomedical Circuits and Systems. 2016;**10**(5):979-989. DOI: 10.1109/TBCAS.2016.2

[28] Tang SC, McDannold NJ. Power loss analysis and comparison of segmented and unsegmented energy coupling coils for wireless energy transfer. IEEE Journal of Emerging and Selected Topics in Power Electronics. 2015;**3**(1): 215-225. DOI: 10.1109/JESTPE. 2014.2330951

[29] Tang SC, Lun TLT, Guo Z, Kwok KW, McDannold NJ. Intermediate range wireless power transfer with segmented coil transmitters for implantable heart pumps. IEEE Transactions on Power Electronics. 2017; **32**(5):3844-3857. DOI: 10.1109/ TPEL.2016.2584558

[30] Stoecklin S, Yousaf A, Gidion G, Reindl L. Efficient wireless power

transfer with capacitively segmented RF coils. IEEE Access. 2020;**8**: 24397-24415. DOI: 10.1109/ ACCESS.2020.2971176

[31] Mark M, Björninen T, Ukkonen L, Sydänheimo L, Rabaey JM. SAR reduction and link optimization for mm-size remotely powered wireless implants using segmented loop antennas. IEEE Topical Conference on Biomedical Wire less Technologies, Networks, and Se nsing Systems. 2011. pp. 7-10. DOI: 10.1109/BIOWIRELESS.2011. 5724339

[32] Pokharel RK, Barakat A, Alshhawy S, et al. Wireless power transfer system rigid to tissue characteristics using metamaterial inspired geometry for biomedical implant applications. Scientific Reports. 2021;**11**:5868

[33] FCC Policy on Human Exposure to Radio Frequency Electromagnetic Fields, https://www.fcc.gov/general/radio-fre quency-safety-0#:~:text=The%20FCC% 20limit%20for%20public,Available% 20at%20FCC%20Web%20Site [20] FCC Policy on Human Exposure to Radio Frequency Electromagnetic Fields, https://www.fcc.gov/general/radio-frequency-safety-0#:~:text=The%20FCC %20limit%20for%20public,Available% 20at%20FCC%20Web%20Site

[34] 1999/519/EC: Council Recommendation of 12 July 1999 on the limitation of exposure of the general public to electromagnetic fields (0 Hz to 300 GHz). https://op.europa.eu/en/pub lication-detail/-/publication/9509b04f-1df 0-4221-bfa2-c7af77975556/language-en#: ~:text=Home-,1999%2F519%2FEC%3A% 20Council%20Recommendation%20of% 2012%20July%201999,0%20Hz%20to% 20300%20GHz) https://op.europa.eu/e n/publication-detail/-/publication/9509b 04f-1df0-4221-bfa2-c7af77975556/lang uage-en#:~:text=Home-,1999%2F519% 2FEC%3A%20Council% 20Recommendation%20of%2012%20July %201999,0%20Hz%20to%20300% 20GHz)

[35] Part 15: Radio Frequency Devices: Code of Federal Regulations, Title 47. https://www.ecfr.gov/current/title-47/chapter-I/subchapter-A/part-15

[36] Part 18: Industrial, Scientific, and Medical Equipment: Code of Federal Regulations, Title 47. https://www.ecfr.g ov/current/title-47/chapter-I/subchapte r-A/part-18

[37] ETSI EN 300 330 V2.1.1 (2017-02) Short Range Devices (SRD); Radio equipment in the frequency range 9 kHz to 25 MHz and inductive loop systems in the frequency range 9 kHz to 30 MHz. https://standards.iteh.ai/catalog/standa rds/etsi/c765eb9a-f7ae-4089-925b-67701bd00d53/etsi-en-300-330-v2.1.1-2017-02

[38] ETSI EN 303 417 V1.1.1 (2017-09) Wireless power transmission systems, using technologies other than radio frequency beam in the 19 - 21 kHz, 59 - 61 kHz, 79 - 90 kHz, 100 - 300 kHz, 6765 - 6795 kHz ranges. https://cdn.sta ndards.iteh.ai/samples/47585/236745461a 87403b994d6e41e4c58861/ETSI-EN-303-417-V1-1-1-2017-09-.pdf

[39] ETSI EN 300 220-2 V3.2.1 (2018-06) Short Range Devices (SRD) operating in the frequency range 25 MHz to 1 000 MHz; Part 2: Harmonised Standard for access to radio spectrum for non specific radio equipment. https://www.etsi.org/ deliver/etsi_en/300200_300299/ 30022002/03.02.01_60/en_ 30022002v030201p.pdf

[40] COMMISSION IMPLEMENTING REGULATION (EU) No 572/2013 of 19 2013. https://eur-lex.europa.eu/

legal-content/EN/TXT/PDF/?uri=CELEX:32013R0572&from=EN

[41] EN IEC 55014-1:2021 Electromagnetic compatibility - Requirements for household appliances, electric tools and similar apparatus - Part 1: Emission. https://standards.iteh.ai/catalog/standards/clc/5da378f9-f8a7-4cd1-84b6-e40d63b355b9/en-iec-55014-1-2021

[42] EN 55011:2016/A11:2020 Industrial, scientific and medical equipment - Radio-frequency disturbance characteristics - Limits and methods of measurement. https://standards.iteh.ai/catalog/standards/clc/179803ac-9c0f-499a-8d92-dabaa9c6f1c7/en-55011-2016-a11-2020